The MIDI Manual

The MIDI Manual: A Practical Guide to MIDI within Modern Music Production, Fourth Edition, is a complete reference on MIDI. Written by David Miles Huber (a 4× Grammy-nominated musician, producer and author), this best-selling guide provides clear explanations of what MIDI 1.0 and 2.0 are, acting as a guide for electronic instruments, the DAW, MIDI sequencing and how to make best use of them. You will learn how to set up an efficient MIDI system and how to get the most out of your production room and ultimately … your music.

Packed full of useful tips and practical examples on sequencing and mixing techniques, *The MIDI Manual* also covers in-depth information on system interconnections, controllers, groove tools, the DAW, synchronization and more. For the first time, the MIDI 2.0 spec is explained in light of the latest developments and is accompanied with helpful guidelines for the long-established MIDI 1.0 spec and its implementation chart.

Illustrated throughout with helpful photos and screenshots, this is the most readable and clearly explained book on MIDI available.

David Miles Huber is a 4× Grammy-nominated producer and musician in the electronic dance and immersive genres, whose music has sold over the million mark. His dance performance style is energized and balanced out by lush beats and live acoustic instruments that combine to create a "Zen-Meets-Tech Experience". His latest music can be heard at www.davidmileshuber.com and www.davidmileshuber.bandcamp.com.

DMH received his degree in music technology from Indiana University and the University of Surrey in Guildford, England. His most prominent book *Modern Recording Techniques*, Ninth Edition, is the standard recording industry text worldwide.

AUDIO ENGINEERING SOCIETY PRESENTS...

www.aes.org

Editorial Board

Women in Audio
Leslie Gaston-Bird

Audio Metering
Measurements, Standards and Practice
Eddy B. Brixen

Classical Recording
A Practical Guide in the Decca Tradition
Caroline Haigh, John Dunkerley and Mark Rogers

The MIDI Manual, Fourth Edition
A Practical Guide to MIDI within Modern Music Production
David Miles Huber

Digital Audio Forensics Fundamentals
From Capture to Courtroom
James Zjalic

For more information about this series, please visit: www.routledge.com/Audio
-Engineering-Society-Presents/book-series/AES

The Midi Manual

A Practical Guide to MIDI within Modern Music Production

Fourth Edition

David Miles Huber

Routledge
Taylor & Francis Group

NEW YORK AND LONDON

Fourth edition published 2021
by Routledge
52 Vanderbilt Avenue, New York, NY 10017

and by Routledge
2 Park Square, Milton Park, Abingdon, Oxon, OX14 4RN

Routledge is an imprint of the Taylor & Francis Group, an informa business

First edition published by Focal Press 2007
Third edition published by Focal Press 2009

British Library Cataloguing-in-Publication Data
A catalogue record for this book is available from the British Library

Library of Congress Cataloging-in-Publication Data
Names: Huber, David Miles, author.
Title: The MIDI manual 4e: a practical guide to MIDI within modern
music production/David Miles Huber.
Description: Fourth edition. | New York: Taylor and Francis, 2021. |
Series: Audio Engineering Society presents | Includes index.
Identifiers: LCCN 2020020249 (print) | LCCN 2020020250 (ebook) |
ISBN 9780367549985 (paperback) | ISBN 9780367549978 (hardback) |
ISBN 9781315670836 (ebook)
Subjects: LCSH: MIDI (Standard) | Software sequencers. |
Sound recordings--Production and direction.
Classification: LCC MT723 .H82 2021 (print) | LCC MT723 (ebook) |
DDC 784.190285/46--dc23
LC record available at https://lccn.loc.gov/2020020249
LC ebook record available at https://lccn.loc.gov/2020020250

ISBN: 978-0-367-54997-8 (hbk)
ISBN: 978-0-367-54998-5 (pbk)
ISBN: 978-1-315-67083-6 (ebk)

Typeset in Giovanni Book
by Deanta Global Publishing Services Chennai India

Contents

v

x | **Contents**

Foreword

The MIDI Manual was first published in 1991. Nearly three decades later, the information in this new edition is still vital to anyone who wants to create music using today's technology.

The Musical Instrument Digital Interface (MIDI) specification was first introduced at the 1983 National Association of Music Merchants (NAMM) Show, with a simple primary goal: play a note on one MIDI hardware keyboard that's physically connected to another MIDI hardware keyboard, and have them both sound at the same time.

Yet this relatively modest specification continued to develop, adapt and expand. Today, you can connect a MIDI controller via wireless Bluetooth LE, and play and record hundreds of notes of software-based synthesis on your smartphone … or program lighting changes in a Broadway play, for that matter.

The key to MIDI's continued evolution is the collaborative effort of an entire industry, from giant multinational corporations to single-person companies. They set aside their differences and competitive nature to embrace a spirit of cooperation and mutual respect. As one MIDI user said, *"MIDI is without a doubt an example of what happens when competing companies work together for common good. In no other industry before or since have we seen something so noble".* MIDI has been connecting people, companies and products for almost 40 years – an unprecedented lifespan for a technical standard that started back when computer memory was measured in kilobytes, not gigabytes.

The MIDI specification's evolution is overseen by the MIDI Manufacturers Association (MMA), a volunteer, non-profit trade organization that works in tandem with AMEI, the Japanese equivalent. With their stewardship, MIDI's capabilities continued to grow after its introduction.

In 1991, the MMA adopted the Standard MIDI File specification, which simplified the sharing of MIDI recordings among users. That year also saw the adoption of MIDI time code, which synchronized MIDI sequencers and tape machines – and revolutionized how Hollywood scored movies.

In the late 1990s, music production started incorporating software-based synthesizers and digital audio workstations for music creation and editing, but you still needed a hardware MIDI controller to input music into your computer. MIDI controllers expanded beyond keyboards to guitars, drums, wind instruments, voice and more. Almost every instrument on the planet, from harmonicas to bagpipes, has been adapted to generate MIDI control signals. When DJs went digital in the early 2000s, MIDI provided the standard interface among products, and because Apple, Google and Microsoft all support MIDI in their operating systems, MIDI is incorporated in literally billions of computers, tablets and smartphones. As music industry legend Craig Anderton says, "MIDI is in the DNA of all modern music productions. Many people don't even realize how much of what they do would be impossible without MIDI, because it has been integrated so transparently, and successfully, into the musical mainstream".

All of these innovations were achieved with the same basic set of messages (MIDI 1.0 protocol) agreed to in 1983. But by 2015, the MIDI 1.0 protocol had reached a dead end, because there was no room left in the spec for further expansion. Yet people clamored for more resolution, more channels, more expressive possibilities, smoother workflow and other advantages made possible by advances in computer technology. Other industries might have simply created a new specification, and said "sorry, your 40 years of equipment is obsolete. You've gotten enough use out of it anyway, just go buy all new gear".

That wasn't an acceptable answer for AMEI and the MMA, and thus began work on improving and enhancing the MIDI spec to meet the needs of not only today, but tomorrow – while remaining backward-compatible with the universe of existing MIDI gear. That work culminated in February 2020, with the adoption of MIDI 2.0. The main breakthrough was turning MIDI from a monolog, where one instrument sent messages to another, to a dialog, where instruments could talk with each other and provide functions such as auto-configuration. This was also the key to backward compatibility – if a MIDI device tried to start a conversation with another MIDI device and didn't get an answer, it would know to fall back to the MIDI 1.0 protocol. MIDI 2.0 gear is now being designed, and the MMA gave David a sneak peak of the specifications so he could include a preview in this new edition. The latest information on MIDI 2.0 is always available from the MMA at www.midi.org/.

But understanding MIDI 2.0 requires a firm background in MIDI 1.0. MIDI 2.0 isn't a completely different language, it's just "more MIDI". So, all of the concepts described in David Huber's *The MIDI Manual* still apply to MIDI 2.0 (and future expansions to come).

You would be hard-pressed to find a better reference to MIDI than *The MIDI Manual*. David not only understands MIDI, but explains it from a user's perspective, because he has been creating and exploring MIDI's creative possibilities for decades. We're sure you will enjoy this book, and it can serve as your MIDI reference guide for many years to come.

Athan Billias
Marketing Technology Strategy Manager
Yamaha Corporation of America

Craig Anderton
Author, Musician and Electronic Music Guru

CHAPTER 1

What Is MIDI?

Simply stated, Musical Instrument Digital Interface (MIDI) is a digital communications language and compatible specification that allows multiple hardware and software electronic instruments, performance controllers, computers and other related devices to communicate with each other over a connected network (Figure 1.1). MIDI is used to translate performance- or control-related events (such as playing a keyboard, selecting a patch number, varying a controller modulation wheel, triggering a staged visual effect, etc.) into equivalent digital messages and then transmit these messages to other MIDI devices where they can be used to make music and/or control performance parameters. The beauty of MIDI is that its data can be recorded into the "sequenced" track of a software Digital Audio Workstation (DAW), where it can then be edited and transmitted to electronic instruments or other devices within a networked system.

In artistic terms, this digital language is an important medium that lets artists express themselves with a degree of flexibility, repeatability and control that, before its inception, simply wasn't possible. Through the transmission of this control language, an electronic musician can create and develop a song or composition in a practical, flexible, affordable and fun production environment that can be tailored to his or her needs.

In addition to composing and performing a song, musicians can also act as creative producers, having complete control over a wide palette of sounds, their timbre (sound and tonal quality), overall blend (level, panning, effects, etc.) and other real-time controls. MIDI can also be used to vary the performance and control parameters of electronic instruments, recording devices, controllers and signal processors in the studio, on the road or on-stage.

The term "interface" refers to the actual data communications link between software/hardware systems within a connected MIDI network. Through MIDI, it's possible for all of the electronic instruments and devices within a network to be addressed using real-time performance and control-related MIDI data messages throughout a system to multiple instruments and devices over one or more data networks. This is possible because a single, connected network is capable of transmitting a wide range of performance- and control-related data, in a way that allows creative individuals to record, overdub, mix and playback their performances in

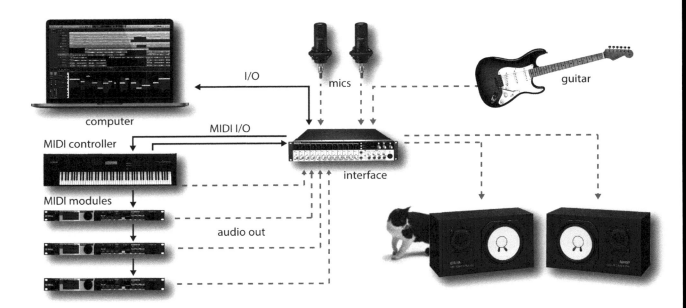

computer
MIDI controller
MIDI modules
I/O
MIDI I/O
audio out
mics
interface
guitar

FIGURE 1.1
Example of a typical MIDI
system with the MIDI network
connections highlighted.

a working environment that loosely resembles the multitrack recording process. Beyond this, however, MIDI surpasses this analogy by allowing a composition to be edited, controlled and altered with complete automation and repeatability – providing a wide range of production possibilities that are well beyond the capabilities of the traditional analog and digital audio process.

THE DIGITAL WORLD

One of the best ways to gain insight into how the MIDI specification works is to compare MIDI to a spoken language. As humans, we've adapted our communication skills to best suit our physical bodies and brains. Ever since the first grunt, we've found that it's easy for us to communicate with our vocal cords … and we've been doing it ever since. Over time, language developed by assigning a standardized meaning to a series of vocalized sounds (words). Eventually these words came to be grouped in such a way as to convey meanings that can be easily communicated … and, finally, written. For example, in order to write down the English language, a standard notation system was developed that assigned 26 symbols to specific sounds (letters of the alphabet) that, when grouped together, could communicate an equivalent spoken word (Figure 1.2). When these words are strung into complete sentences, a more complex form of communication is used that conveys information in a way that has a greater meaning. For example, the letters B, O, O and K don't mean much when used individually; however, when grouped into a word, they refer to a physical media form that hopefully conveys messages relating to a general theme. Changing a symbol in the word could change its meaning entirely; for example, changing the K to a T within the grouped word makes it refer to a handy thing that's better

b, o, o, k =   (0100 0010)(0100 1111)(0100 1111) (0100 1010) =

(alpha-bits)   (digital words)   (book)

FIGURE 1.2
Meaning is given to the alphabet letters "b, o, o and k" when they're grouped into a word or placed into a sentence.

off worn on your feet than carried in a backpack (e.g., Dude, is that a book about pirate boots?).

Microprocessors and computers, on the other hand, are digital devices that obviously lack vocal cords and ears (although even that's changing). However, since they have the unique advantage of being able to process numbers at a very high rate, digital is the obvious language for communicating information at high speeds with complete repeatability.

Unlike our base-10 system of counting, computers are limited to communicating with a binary system of 0s and 1s (off and on). Like humans, computers group these binary digits (known as bits) into larger numeric "words" that represent and communicate specific information and instructions. Just as humans communicate using simple sentences, a computer can generate and respond to a series of related digital words that are understood by other digital hard- and software systems (Figure 1.3).

WHAT MIDI ISN'T

For starters, let's dispel one of MIDI's greatest myths: MIDI does not communicate audio nor can it create sounds on its own! It is nothing more or less than a digital language that instructs a device or program to create, playback or alter

FIGURE 1.3
Example of a digitally generated MIDI message.

MIDI interface

in
out ⟶ (1101 CCCC) (NNNN NNNN) (VVVV VVVV) (1101 CCCC) (NNNN NNNN) (VVVV VVVV)

in out thru

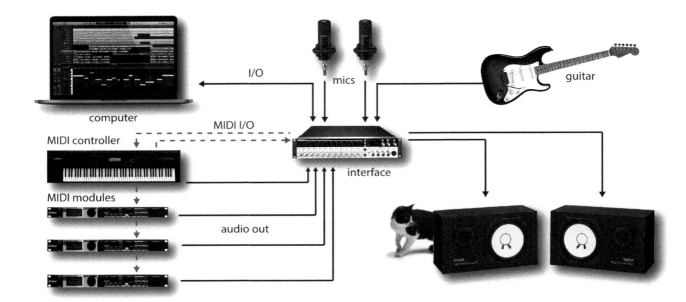

FIGURE 1.4
Example of a typical MIDI system with the audio connections highlighted.

sounds. MIDI is a data protocol that communicates control info and parameter messages to instruct instruments and/or devices to generate, reproduce or control production-related functions. Because of these differences, the MIDI data path and the audio routing paths are entirely separate from each another (Figure 1.4). Even if they digitally share the same transmission cable (such as through USB, Thunderbolt, FireWire or LAN), the actual data paths and formats will remain completely separate.

Put another way, MIDI communicates information that instructs a digital instrument or device to play or carry out a function. It can be thought of as the dots on an old player-piano roll. When we put the paper roll up to our ears, we hear nothing, but when the cut-out dots pass over the sensors on a player piano, the instrument itself begins to make music. It's exactly the same with MIDI. A MIDI file or data stream is simply a set of instructions that pass down a wire in a serial fashion, but when an electronic instrument interprets the data … only then can we begin to hear sound.

WHAT MIDI IS

In everyday use, MIDI can be thought of as a compositional tool for capturing, editing and controlling production-related media. It's an amazingly powerful environment that, as with the digital world, is extremely chameleon-like in nature.

- It can be used in straightforward ways, allowing sounds and textures to be created, edited, mixed and blended into a composition.

- It can be used in conjunction with groove and looping tools to augment, control and shape a production in an endless number of ways and over a wide range of music genres.
- It is a tool for capturing a live performance (as a tip, if an instrument in the studio has a MIDI out jack, it's always wise to record it to a MIDI track on your DAW). The ability to edit, change a sound or vary parameters after the fact is a useful tool that could save, augment and/or improve a performance.
- MIDI, by its very nature, is a "re-amp" beast; the ability to change a sound, instruments, settings and/or parameters in post-production is largely what MIDI is all about. You could even play an instrument back in the studio, turn it up and capture the electronically generated sounds acoustically in the studio or bedroom by using mics … there are practically no limits to your creative options.
- The ability to have real-time and post-production control over music and effects parameters is literally in MIDI's DNA. Almost every parameter can be mangled, mutilated and finessed to fit your wildest dreams – either during the composition phase or in post-production.

In short, the name of this game is editability, flexibility and individuality. There are so many ways of approaching and working with MIDI that there are few wrong ways to approach a project … it's very individualistic in its nature. Because of the ways that a system can be set up, the various approaches by which the tools and toys are used to create music and sounds are very personal. How you use your tools to create your own style of music is literally up to you, both in production and in post-production… That's the true beauty of MIDI.

MIDI 1.0 – A BRIEF HISTORY

In the early days of electronic music (Figure 1.5), keyboard synthesizers were commonly only monophonic devices (capable of sounding only one note at a time) and often generated tones that were often simple in nature. These limiting

FIGURE 1.5
The late, great synth pioneer Bob Moog, who was outstanding in his field. (Photograph courtesy of Roger Luther; www.moogarchives.com.)

factors caused early manufacturers to look for ways to combine instruments together to create a thicker, richer sound texture (known as layering). This was originally accomplished by establishing an instrument link that would allow a synthesizer (acting as a master controller) to directly control the performance parameters of one or more other synthesizers (known as slave sound modules). As a result of these links, a basic control signal (known as control voltage or CV) was developed.

This simple yet problematic system was based on the fact that when most early keyboards were played, they generated a DC voltage that could directly control another instrument's voltage-controlled oscillators (VCO – which affected the pitch of a sounding note) and voltage-controlled amplifiers (VCA – which affected the note's volume and on/off nature). Since many keyboards of the day generated a DC signal that ascended at a rate of 1 volt per octave (breaking each musical octave into 1/12-volt intervals), it was possible to use this standard control voltage as a master-reference signal for transmitting pitch information to other synths. In addition to a control voltage, this standard required that a keyboard transmit a gate signal. This second signal was used to synchronize the beginning and duration times of each note. With the appearance of more advanced polyphonic synthesizers (which could generate more than one note at a time) and early digital devices, it was clear that this standard would no longer be the answer to properly controlling a connected system. Thus, new standards began to appear on the scene, thereby creating the crazy fun of having incompatible control standards. With the arrival of early drum machines and sequencing devices, standardization became even more of a dilemma.

Synchronization between these early devices was also problematic, as manufacturers would often standardize on different sync-pulse clock rates. Synchronizing incompatible systems could be extremely difficult, because they would lose their timing over a very short period of time, rendering sync nearly impossible without additional sync-rate converters or other types of modifications. Because of this mess, Dave Smith and Chet Wood (then of Sequential Circuits, a now-defunct manufacturer of electronic instruments) began creating a digital electronic instrument protocol, which was named the Universal Synthesizer Interface (USI). As a result of this early protocol, equipment from different manufacturers could finally communicate directly (e.g., a synth from one company finally worked with another company's sequencer). In the fall of 1981, USI was proposed to the Audio Engineering Society. During the following two years, a panel (which included representatives from the major electronic instrument manufacturers) modified this standard and adopted it under the name of Musical Instrument Digital Interface … and the MIDI Specification 1.0 was born.

The strong acceptance of MIDI was largely due to the need for a standardized protocol, as well as the fast-paced advances in technology that allowed complex circuit chips and hardware designs to be manufactured cost-effectively. It was also due, in part, to the introduction of Yamaha's wildly popular DX-7 synthesizer in the winter of 1983, after which time keyboard sales began to grow at an astonishing rate.

With the adoption of this industry-wide standard, any device that incorporated MIDI into its design could transmit or respond to digital performance and control-related data that conformed to the MIDI 1.0 spec. For the first time, any new device that conformed to this spec would integrate into any existing MIDI system and actually work … without any muss or fuss.

Over the course of time, new devices came onto the market that offered improved sound and functional capabilities that led to the beginnings of hardware/software instruments, controllers, effects and a whole world of gizmos that could be controlled in the MIDI environment. In fact, this explosion of software emulation and control has breathed a new degree of life into the common, everyday use of this amazing medium.

MIDI 2.0 – A MORE RECENT HISTORY

For over 30 years, the MIDI 1.0 spec remained unchanged. This is a true testament to the hard work and dedication that the original members of the MIDI Manufacturers Association (MMA, www.midi.org) and the Association of Music Electronics Industry (AMEI, its Japanese counterpart) put into making a spec that would withstand the test of time. However, at the 2019 winter National Association of Music Merchants (NAMM) Convention, it was announced that the spec was getting an overhaul and, in the fall of 2019, MIDI 2.0 was ratified for general industry use.

Within Chapter 3, we'll be taking an in-depth look into the changes and enhancements that went into the creation of MIDI 2.0, including bi-directionality, increased performance and controller resolution, expanded feature-sets, increased port/channel capabilities and so forth. Above all, the MMA and AMEI have made backwards compatibility a key requirement, stating that users can expect MIDI 2.0 and the related newer systems to work seamlessly with MIDI devices that have been sold over the past 33 years.

WHY IS MIDI?

As we read, before MIDI, it was pretty much necessary to perform a musical passage in real time. Of course, there are a few exceptions to this statement. In earlier days, sounds could be created and re-created though the mechanical triggering of a musical device (music boxes and the aforementioned player piano come to mind). When tape-based recording came along in the middle part of the last century, it became possible to edit two or more problematic performances together into a single, good "take". However, when it came to the encoding of a musical passage and then faithfully playing it back – while still being able to edit or alter the tempo, notes and control variables of a performance – we were pretty much back in the horse-and-buggy days.

With the introduction of MIDI and electronic music production, a musical performance could be captured in the digital domain and then faithfully played back in a production-type environment that mimicked the traditional form and functions of multitrack recording. Basic tracks could be recorded one at a time, allowing a composition to be built up using various electronic instruments. But, here's the kicker: MIDI finally made it possible for a performance track to be edited, layered, altered, spindled, mutilated and (most of all) faithfully repeated with relative ease, under completely automated computer control. If you played a bad note … you can simply fix it. If you want to change the key or tempo of a piece, just change it. If you want to change the expressive volume of a phrase in a song, just do it! Want to change the piano patch (sound) into a wildly gyrating synth … no problem. These editing capabilities only hint at the full power of MIDI!

This affordable potential for future expansion and increased control throughout an integrated production system has spawned the growth of an industry that's also very personal in nature. For the first time in music history, it's possible for an individual to cost-effectively realize a full-scale sound production in their own time … all within the comfort of one's own home or personal project studio.

I'd also like to address another issue that has sprung up around MIDI and electronic music production. With the introduction of drum machines, modern-day synths, samplers and powerful hardware/software instruments, it's not only possible but it is also relatively easy to build up a composition using instrument voices that closely mimic virtually any instrument that can be imagined. In the early days, studio musicians spoke out against MIDI, saying that it would be the robot that would make them obsolete. Although there was a bit of truth to this, these same musicians are now using the power of MIDI to expand their own musical palate and create productions of their own. Today, MIDI is being used by professional and nonprofessional musicians alike to perform an expanding range of production tasks, including music production, audio-for-video and film post-production and live on-stage production. Such is progress in the modern world.

MIDI IN THE HOME STUDIO

It almost goes without saying that a vast number of electronic musical instruments, effects devices, computer systems and other MIDI-related devices are currently available on the new and used electronic music market. With the introduction of the large-scale integrated circuit chip (which allowed complex circuitry to be quickly and easily mass produced), many of the devices that make up an electronic music system have been made affordable to virtually every musician or composer, whether he or she is a working professional, aspiring artist or beginning hobbyist (Figure 1.6). This amazing diversity lets us build up a home project studio that's very personal in nature, allowing us to create an environment that best suits our own particular musical taste and production style … all in the comfort of our own home.

MIDI ON THE GO

Of course, MIDI production systems can appear in any number of shapes and sizes and can be designed to match any production and budget need. For example, these days, on-the-go musicians can often be seen pulling out their laptops, portable keyboard and headphone/ear buds to lay down a track on a plane, bus or simply laying by the pool with a margarita at hand. Even more portable toys, such as the iPad, might be the tool of choice for getting the job done while you're drifting down a slow river in a small boat. Obviously, these small systems have gotten powerful enough that they can let us compose, produce and mix virtually anywhere (Figure 1.7).

MIDI IN THE STUDIO

MIDI has also dramatically changed the sound, technology and production habits of the recording studio. Before MIDI and the concept of the home project studio, the professional recording studio was pretty much the only place that allowed an artist or composer to combine instruments and sound textures into a final recorded product. Often, the process of recording a group in a live setting

FIGURE 1.6
Gettin' it all going in the project studio. (Courtesy of Yamaha Corporation of America; www.yamaha.com.)

FIGURE 1.7
Between takes. (Photograph courtesy of M-Audio, a registered trademark of inMusicBrands, LLC; www.m-audio.com.)

FIGURE 1.8
MIDI in the studio. (Courtesy
of Steinberg Media
Technologies GmbH, a division of Yamaha Corporation,
www.steinberg.net.)

was (and still can be) an expensive and time-consuming process. This is due to the high cost of hiring session musicians and the hourly rates that are charged for a professional studio – not to mention Murphy's studio law, which states that you'll always spend more time and money than you thought you would in an effort to capture that elusive "ideal performance".

With the advent of the DAW and MIDI, much of the music production process can now be preplanned and rehearsed (if not totally produced and recorded) before you step into the studio (Figure 1.8). This out-and-out luxury has drastically reduced the number of hours that are needed for laying down recorded tracks to a cost-effective minimum. We're all aware that it's often the norm for musicians to record and produce entire albums in their own project studios. If the artist doesn't feel that they have the expertise to see the process through to its final form (it's often a wise person who knows their limitations), the artist(s) could print the various instrument tracks into the DAW and then bring the session files into a professional studio (obviously, a much easier process than lugging all of their equipment around).

In essence, through the use of careful planning and preproduction in the home or project studio, a project can be produced in a pro-environment in a much more timely fashion (and hopefully on budget) than would otherwise be possible.

MIDI IN AUDIO-FOR-VISUAL AND FILM

Electronic music has long been an indispensable tool for the scoring of full-feature motion picture sound tracks, as well as in the production and post-production of television commercials, industrial videos, television shows and the like (Figure 1.9). For productions that are on a tight budget, entire scores will often be written and produced in a project studio at a mere fraction of what it might cost to hire musicians and a studio.

Even mega-budget projects make extensive use of MIDI in their pre-production and production phases, as it's the norm that sound cues and orchestral scores

for such projects to be composed, edited and finessed into a MIDI version of the composer's score before the final score can be printed and distributed to the musicians at the final scoring session.

FIGURE 1.9
MIDI for visual media can be made a major production facility or it can be produced just as easily in a home production environment.

MIDI IN LIVE PERFORMANCE

Electronic music production and MIDI are also very much at home on the stage. Obviously, MIDI has played a crucial role in helping to bring live music to the masses. The ability to sequence rhythm and background parts in advance, chain them together into a single, controllable sequence (using a jukebox-type sequencing program) and then play them on stage has become an indispensable live-performance tool for many working musicians. This technique is widely used by solo artists who have become one-man bands by singing and playing their guitar to a series of background sequences.

Again, the power of MIDI lies in the fact that much of a performance can be composed and produced before going on stage or on tour. Also, MIDI can be used in on-the-fly improvisation, adding a fresh and varied feel to the performance for both those on stage and in the audience. In addition to communicating performance data, MIDI controllers can provide a wide range of control over performance parameters in real time.

In addition to the musical performance, MIDI will often play a strong role in the production and execution of on-stage lighting and special effects. Most modern-day lighting boards are equipped with a MIDI interface, allowing full control over lighting from a sequencer, loop-based controller or MIDI control surface in a way that can pause, vary and execute scene changes in a more interactive and on-the-spot manner (while still allowing scenes to be solidly synchronized to the basic script, when needed).

The ability to offer control over a pre-programmed sequence or interactive video loops has put MIDI directly into the driver's seat when it comes to on-stage and venue visuals and video playback. Many bars and music acts are beginning to integrate visuals into their overall presence – so much so that VJs now stand alongside their bandmates on-stage, offering up compelling visuals that can be diced, sequenced and scratched in forms that instantly switch from being totally chaotic to being in perfect sync and then back again.

MIDI AND MULTIMEDIA

One of the "media" in multimedia is definitely MIDI. It often pops up in places that you might expect – and in others that might take you by complete surprise. With the advent of General MIDI (a standardized specification that makes it possible for any soundcard or GM-compatible device to play back a score using the originally intended sounds and program settings), it's possible (and common) for MIDI scores to be integrated into multimedia games, text documents, CD-ROMs and even websites. Due to the fact that MIDI is simply a series of performance commands (unlike digital audio, which actually encodes the waveform data), the media's data overhead requirements are extremely low. This means that almost no processing power is required to play MIDI, making it the ideal medium for playing real-time music scores while you're actively play a game or browse text, graphics or other media over the Internet. Truly, when it comes to weaving MIDI into the various media types, the sky (and your imagination) is the creative and technological limit.

MIDI ON THE PHONE

With the integration of the Mobile Downloadable Sounds (DLS) Spec into almost all mobile media devices, one of the fastest-growing General MIDI applications is probably comfortably resting in your pocket or purse right now. We're talking about the synthesized music, game sounds and ringtones on your cell phone. The ability to use MIDI (alongside digital sound) to let you know who is calling has spawned an industry that allows your cell to be personalized and super fun. One of my favorite ring tone stories happened to me on Hollywood Boulevard in LA. This tall, lanky man was sitting in a café when his cell phone started blaring out the "If I Only Had a Brain" sequence from *The Wizard of Oz*. It wouldn't have been nearly as funny if the guy didn't look "a lot" like the scarecrow character. Of course, everyone broke out laughing.

CHAPTER 2

The MIDI 1.0 Spec

The Musical Instrument Digital Interface is a digital communications protocol. That's to say, it is a standardized control language and hardware specification that makes it possible for electronic instruments, processors, controllers and other device types to communicate performance and control-related data in real time. Founded in 1983 by a small group from various manufacturers, the MIDI 1.0 protocol has been instrumental in making it possible for all almost all electronic instruments to seamlessly communicate over a connected network. As the MIDI 1.0 spec is fully implemented within the 2.0 spec, it's important that it be covered in its entirety.

MIDI 1.0 is a specified data format (and a complete sub-set protocol within MIDI 2.0) that must be strictly adhered to by those who design and manufacture MIDI-equipped instruments and devices. Because the format is standardized, you don't have to worry about whether the MIDI output of one device will be understood by the MIDI in port of a device that's made by another manufacturer. As long as the data ports say and/or communicate MIDI, you can be assured that the data (at least the basic performance functions) will be transmitted and understood by all devices within the connected system. In this way, the user need only consider the day-to-day dealings that go hand-in-hand with using electronic instruments, without having to be concerned with data compatibility between devices.

THE MIDI MESSAGE

MIDI digitally communicates musical performance data between devices as a string of MIDI messages. These messages are traditionally transmitted through a standard MIDI line in a serial fashion at a speed of 31,250 bits/sec. Within a serial data transmission line, data travel in a single-file fashion through a single conductor cable (Figure 2.1a); a parallel data connection, on the other hand, is able to simultaneously transmit digital bits in a synchronous fashion over a number of wires (Figure 2.1b).

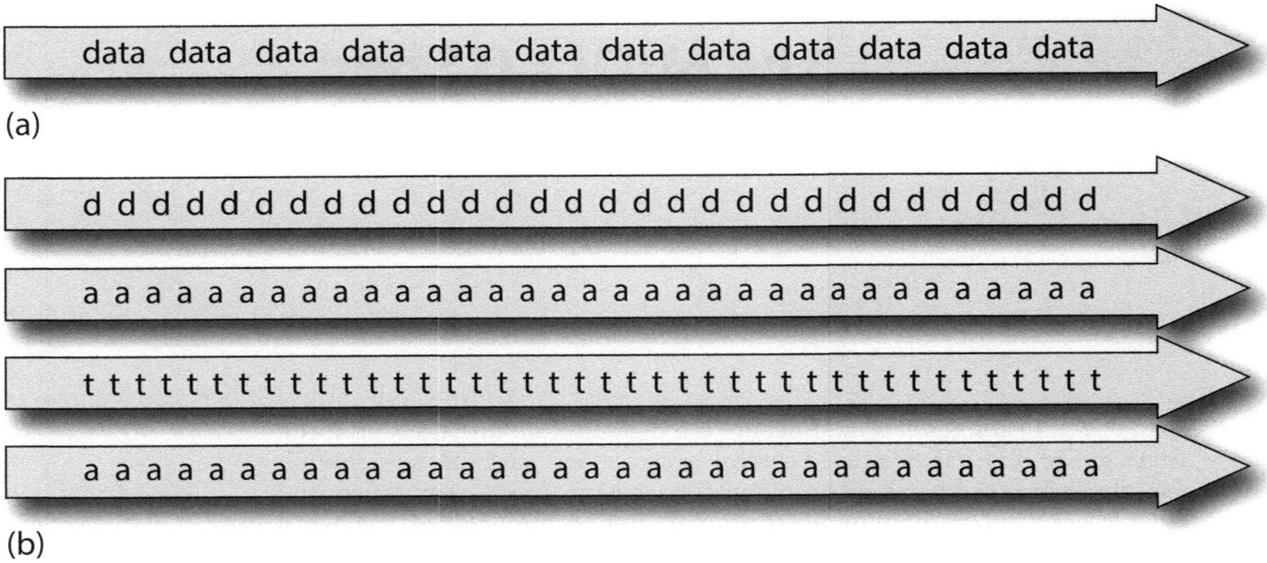

(a)

(b)

FIGURE 2.1
Serial versus parallel data
transmission: (a) serial data
must be transmitted in a
single-file fashion over a
serial data line; (b) multiple
bits of data can be synchronously transmitted over a
number of parallel lines.

When using a standard MIDI 1.0 cable, it's important to remember that data can only travel in one direction from a single source to a destination (Figure 2.2a). In order to make two-way communication possible, a second MIDI data line must be used to communicate data back to the device, either directly or thru the MIDI chain (Figure 2.2b).

MIDI messages are made up of groups of 8-bit words (known as bytes), which are transmitted in a serial fashion to convey a series of instructions to one or all MIDI devices within a system.

Only two types of bytes are defined by the MIDI 1.0 specification: the *status byte* and the *data byte*.

- A status byte is used to identify what type of MIDI function is to be performed by a device or program. It is also used to encode channel data (allowing the instruction to be received by a device that's set to respond to the selected channel).
- A data byte is used to associate a value to the event that's given by the accompanying status byte.

Although a byte is made up of 8 bits, the most significant bit (MSB; the leftmost binary bit within a digital word) is used solely to identify the byte type. The MSB of a status byte is always 1, while the MSB of a data byte is always 0 (Figure 2.3). For example, a 3-byte MIDI Note-On message (which is used to signal the beginning of a MIDI note) in binary form might read as shown in Table 2.1. Thus, a 3-byte Note-On message of (10010100) (01000000) (01011001) will transmit instructions that would be read as: "Transmitting a

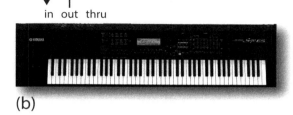

in out thru · in out thru

(a)

in out thru · in out thru

(b)

Note-On message over MIDI channel #5, using keynote #64, with an attack velocity [volume level of a note] of 89".

FIGURE 2.2
MIDI 1.0 data can only travel in one direction through a single MIDI cable: (a) data transmission from a single source to a destination; (b) two-way data communication using two cables.

MIDI Channels

Just as a public speaker might single out and communicate a message to one individual in a crowd, MIDI messages can be directed to communicate information to a specific device or range of devices within a MIDI system. This is done by imbedding a channel-related nibble (4 bits) within the status/channel number byte (Figure 2.4). This makes it possible for performance or control information to be communicated to a specific device – or a sound generator within a device – that's assigned to a particular channel.

Since this nibble is 4 bits wide, up to 16 discrete MIDI channels can be transmitted through a single MIDI cable or designated port.

Whenever a MIDI device, instrument or program function is instructed to respond to a specific channel number, it will only respond to messages that are transmitted on that channel (i.e., it ignores channel messages that are transmitted on any other channel). For example, let's assume that we're going to create a short song using a synthesizer that has a built-in sequencer (a device or program that's capable of recording, editing and playing back MIDI data) and two other instruments (Figure 2.5):

FIGURE 2.3
The most significant bit of a MIDI data byte is used to distinguish between a status byte (where MSB = 1) and a data byte (where MSB = 0).

The MSB of a Status Byte is always "1"
(1SSS SSSS)

The MSB of a Data Byte is always "0"
(0DDD DDDD)

Table 2.1	Status and Data Byte Interpretation		
	Status Byte	**Data Byte 1**	**Data Byte 2**
Description	Status/channel #	Note #	Attack velocity
Binary data	(1001.0100)	(0100.0000)	(0101.1001)
Numeric value	(Note-On/CH#5)	(64)	(89)

DIY: Assigning MIDI Channels

1. We could start off by recording a drum track into the master synth using channel 10 (many synths are pre-assigned to output drum/percussion sounds on this channel).

2. Once recorded, the sequence will transmit the notes and data over channel 10, allowing the synth's internal percussion section to played.

3. Next, we could set another synth module to channel 3 and instruct the master synth to transmit on this same channel (since the synth module is set to respond to data on channel 3. Its generators should now sound whenever the master keyboard is played and we can now begin recording a melody line into the sequencer's next track.

4. Playing back the sequence will then transmit data to both the master synth (percussion section) and the module (melody line) over their respective channels. At this point, our song is beginning to take shape.

5. Now we can set a sampler (or other instrument type) to respond to channel 5 and instruct the master synth to transmit on the same channel, allowing us to further embellish the song.

6. When a song begin to take shape, the sequencer can play the musical parts to the synths on their respective MIDI channels – all in an environment that allows us to have complete control over volume and a wide range of functions over each instrument voice. In short, we've created a true multichannel working environment.

FIGURE 2.4
The least significant nibble (4 bits) of the status/channel number byte is used to encode channel number data, allowing up to 16 discrete MIDI channels to be transmitted through a single MIDI cable or designated port.

It goes without saying that the above example is just but one of the infinite setup and channel possibilities that can be encountered in a production environment. It's often true, however, that even the most complex production rooms will have their own basic channel and overall layout that makes the day-to-day operation of making music easier. This layout and the basic decisions are, of course, up to you. Streamlining a system to work both efficiently and easily will come with time, experience and practice.

The last 4-bit status byte is used to encode the channel number 1 - 16
↓

(1SSS CCCC)

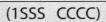

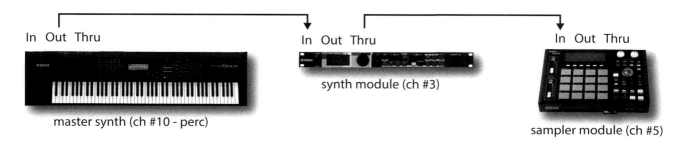

In Out Thru   In Out Thru   In Out Thru

synth module (ch #3)

master synth (ch #10 - perc)

sampler module (ch #5)

Auto-Channelizing

Keeping track of the channels that are assigned to the various instruments and/ or plugins within a session can sometimes be a pain. As the previous paragraph suggests, it's only natural that you'll eventually come up with your own system for assigning MIDI channels to the various devices in your production studio. One of the ways to ease the pain of assigning channels throughout the studio is through the use of an auto-channelizing feature that's built into most sequencers (Figure 2.6). In short, channelizing allows a sequencer to accept incoming data, regardless of its MIDI channel number, and then reassign the outgoing data to the selected track's channel … meaning that the correct MIDI channel will always be outputted to the appropriate device without having to think about it or do anything. This can be best understood by trying it out for yourself:

FIGURE 2.5
MIDI setup showing a set of MIDI channel assignments.

DIY: Auto-Channelizing

1. Plug a keyboard controller into a DAW or MIDI sequencer and make it the active MIDI input source.
2. Create a MIDI track and assign its channel and port numbers to one of your favorite devices (e.g., transmitting on channel 1 to your favorite synth).
3. Create another MIDI track and assign its channel and port numbers to another of your favorite devices (e.g., transmitting on channel 2 on another synth).
4. Create yet another MIDI track and assign its channel and port numbers to another of your favorite devices (e.g., transmitting on channel 10 on yet another synth lead.
5. Next, place the first track into the Record Ready mode and play your keyboard. This

should cause the MIDI data to route to channel 1, allowing the first device to begin playing.
6. Now, unselect the first track and place the second MIDI track into Record/Monitor Ready mode. This should cause the MIDI data to automatically route to channel 2, allowing the second device to begin playing.
7. Finally, place the third track into the Record Ready mode and play your keyboard. Does the output channel routing automatically change so the third device responds to the MIDI data by playing? If not, be patient. Read through the manual and check your cables. If at first you don't succeed…

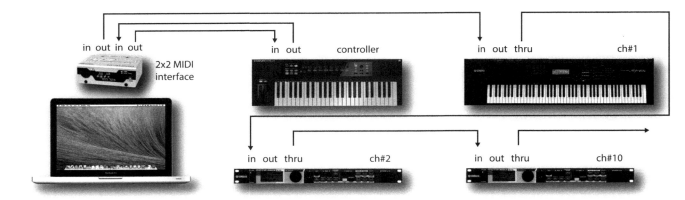

FIGURE 2.6
MIDI setup that makes use of a MIDI controller whereby the MIDI out assignments can be automatically selected through a DAW's enabled MIDI track.

From this, you can see that the MIDI channel or port will automatically change to match that of the selected track – in effect, making MIDI channel assignment much easier and allowing you to stick to the task of making music.

MIDI Modes

Electronic instruments often vary in the number of sounds and notes that can be simultaneously produced by their internal sound-generating circuitry. For example, certain instruments can only produce one note at a single time (known as a *monophonic* instrument), while others can generate 16, 32, and even 64 notes at once (these are known as *polyphonic* instruments). The latter type can easily play chords and/or more than one musical line on a single instrument at a time.

In addition, some instruments are only capable of producing a single generated sound patch (often referred to as a "voice") at any one time. Its generating circuitry could be polyphonic, allowing the player to lay down chords, but it can only produce these notes using a single, characteristic sound at any one time (e.g., an electric piano or a string patch). However, the vast majority of newer synths differ from this in that they're *multi-timbral* in nature, meaning that they can generate numerous sound patches at any one time (e.g., an electric piano, a synth bass, and a string patch, as can be seen in Figure 2.7). That's to say that it is common to run across electronic instruments that can simultaneously generate a number of voices, each offering its own control over parameters (such as volume, panning or modulation). Best of all, it is also common for different sounds to be assigned to their own MIDI channels, allowing multiple patches to be internally mixed within the device to a stereo output bus or independent outputs.

FIGURE 2.7
Multi-timbral instruments are virtual bands-in-a-box that can simultaneously generate multiple patches, each of which can be assigned to its own MIDI channel.

Ch #01	Big Bass
Ch #02	Sub Stick
Ch #03	Brassman
Ch #04	Dyno Pad
....	
Ch #16	Ice Vibes

internal mixer

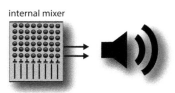

It should be noted that the word "patch" is a direct reference from earlier analog synthesizers, where patch chords were used to connect one sound generator block or processing function to another to create a generated sound.

As a result of these differences between instruments and devices, a defined set of guidelines (known as MIDI reception modes) has been specified that allows a MIDI instrument to transmit or respond to MIDI channel messages in several ways. For example, one instrument might be programmed to respond to all 16 MIDI channels at one time, while another might be polyphonic in nature, with each voice being programmed to respond only to a single MIDI channel.

POLY/MONO

An instrument or device can be set to respond to MIDI data in either the poly mode or the mono mode. Stated simply, an instrument that's set to respond to MIDI data polyphonically will be able to play more than one note at a time. Conversely, an instrument that's set to respond to MIDI data monophonically will only be able to play a single note at any one time.

OMNI ON/OFF

Omni On/Off refers to how a MIDI instrument will respond to MIDI messages at its input. When Omni is turned on, the MIDI device will respond to all channel messages that are being received regardless of its MIDI channel assignment. When Omni is turned off, the device will respond only to a single MIDI channel or set of assigned channels (in the case of a multi-timbral instrument).

The following list and figures explain the four modes that are supported by the MIDI spec in more detail:

- Mode 1-Omni On/Poly – in this mode, an instrument will respond to data that's being received on any MIDI channel and then redirect this data to the instrument's base channel (Figure 2.8a). In essence, the device will play back everything that's presented at its input in a polyphonic fashion, regardless of the incoming channel designations. As you might guess, this mode is rarely used.
- Mode 2-Omni On/Mono – as in Mode 1, an instrument will respond to all data that'sbeing received at its input, without regard to channel designations; however, this device will only be able to play one note at a time (Figure 2.8b). Mode 2 is used even more rarely than Mode 1, as the device can't discriminate channel designations and can only play one note at a time.
- Mode 3-Omni Off/Poly – in this mode, an instrument will only respond to data that match its assigned base channel in a polyphonic fashion (Figure 2.8c). Data that's assigned to any other channel will be ignored. This mode is by far the most commonly used, as it allows the voices within a multi-timbral instrument to be individually controlled by messages that are being received on their assigned MIDI channels. For example, each of

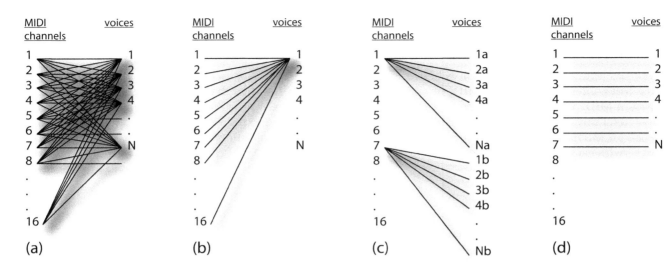

FIGURE 2.8
Voice/channel assignments
of the four modes that are
supported by the MIDI spec:
(a) Omni On/Poly; (b) Omni
On/Mono; (c) Omni Off/Poly;
(d) Omni Off/Mono.

the 16 channels in a MIDI line could be used to independently play each of the parts in a 16-voice, multi-timbral synth.

• Mode 4-Omni Off/Mono – as with Mode 3, an instrument will be able to respond to performance data that's transmitted over a single, dedicated channel; however, each voice will only be able to generate one MIDI note at a time (Figure 2.8d). A practical example of this mode is often used in MIDI guitar systems, where MIDI data is monophonically transmitted over six consecutive channels (one channel/voice per string).

BASE CHANNEL

The assigned base channel determines which MIDI channel the instrument or device will respond to. For example, if a device's base channel is set to 1, it will respond to performance messages arriving on channel 1 while ignoring those arriving on channels 2 through 16.

Channel Voice Messages

Channel Voice messages are used to transmit real-time performance data throughout a connected MIDI system. They're generated whenever a MIDI instrument's controller is played, selected or varied by the performer. Examples of such control changes could be the playing of a keyboard, pressing of program selection buttons or movement of modulation or pitch wheels. Each Channel Voice message contains a MIDI channel number within its status byte, meaning that only devices that are assigned to the same channel number will respond to these commands. There are seven Channel Voice message types: Note-On, Note-Off, Polyphonic Key Pressure, Channel Pressure, Program Change, Pitch Bend Change and Control Change.

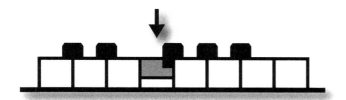

Status / Ch #	Note #	Attack Velocity
(0 - 15)	(0 - 127)	(0 - 127)
(1001 CCCC)	(NNNN NNNN)	(VVVV VVVV)

NOTE-ON MESSAGES

A Note-On message is used to indicate the beginning of a MIDI note. It is generated each time a note is triggered on a keyboard, controller or other MIDI instrument (i.e., by pressing a key, hitting a drum pad or by playing a sequence). A Note-On message consists of 3 bytes of information (Figure 2.9):

- Note-On status/MIDI channel number
- MIDI pitch number
- Attack velocity value

The first byte in the message specifies a Note-On event and a MIDI channel (1–16). The second byte is used to specify which of the possible 128 notes (numbered 0–127) will be sounded by an instrument. In general, MIDI note number 60 is assigned to the middle C key of an equally tempered keyboard, while notes 21 to 108 correspond to the 88 keys of an extended keyboard controller. The final byte is used to indicate the velocity or speed at which the key was pressed (over a value range that varies from 0 to 127). Velocity (Figure 2.10) is used to denote the loudness of a sounding note, which increases in volume with higher velocity values (although velocity can also be programmed to work in conjunction with other parameters, such as expression, control over timbre, or sample voice assignments).

Not all instruments are designed to interpret the entire range of velocity values (as with certain drum machines), and others don't respond dynamically at all. Instruments that don't support velocity information will generally transmit an attack velocity value of 64 for every note that's played, regardless of the how soft or hard the keys are actually being pressed. Similarly, instruments that don't respond to velocity messages will interpret all MIDI velocities as having a value of 64.

FIGURE 2.9
Byte structure of a MIDI Note-On message.

FIGURE 2.10
Velocity is used to communicate the volume or loudness of a note within a performance.

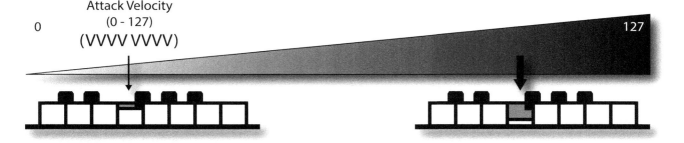

A Note-On message that contains an attack velocity of 0 (zero) is generally equivalent to the transmission of a Note-Off message. This common implementation tells the device to silence a currently sounding note by playing it with a velocity (volume) level of 0.

NOTE-OFF MESSAGES

A Note-Off message is used as a command to stop playing a specific MIDI note. Each Note-On message will continue to play until a corresponding Note-Off message for that note has been received.

In this way, the bare basics of a musical composition can be encoded as a series of MIDI Note-On and Note-Off events. It should also be pointed out that a Note-Off message won't cut off a sound; it'll merely stop playing it. If the patch being played has a release (or final decay) slope, it will begin this stage upon receiving the message. As with the Note-On message, the Note-Off structure consists of 3 bytes of information:

- Note-Off status/MIDI channel number
- MIDI note number
- Release velocity value

In contrast to the dynamics of attack velocity, the release velocity value (0–127) indicates the velocity or speed at which the key was released. A low value indicates that the key was released very slowly, whereas a high value shows that the key was released quickly. Although not all instruments generate or respond to MIDI's release velocity feature, instruments that are capable of responding to these values can be programmed to vary a note's speed of decay, often reducing the signal's decay time as the release velocity value is increased.

ALL NOTES OFF

On the odd occasion (often when you least expect it), a MIDI note can get stuck! This can happen when data drop out or a cable gets disconnected, creating a situation where a note receives a Note-On message but not a Note-Off message, resulting in a note that continues to plaaaaaaaaaaayyyyyyyyyyy! Since you're often too annoyed or under pressure to take the time to track down which note is the offending sucka, it's generally far easier to transmit an All Notes Off message that silences everything on all channels and ports. If you get stuck in such a situation, pressing the "panic button" that "might" be built into your sequencer or hardware MIDI devices could save your ears and sanity.

Pressure (AfterTouch) Messages

Pressure messages (often referred to as "*AfterTouch*") occur after you've pressed a key and then decide to press down harder on it. For devices that can respond to (and therefore generally transmit) these messages, AfterTouch can often be

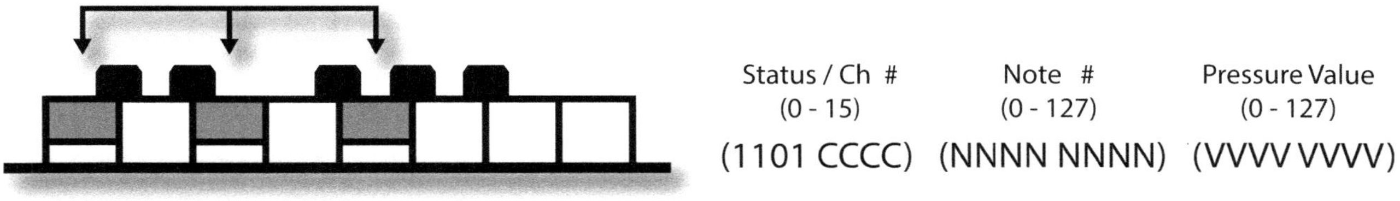

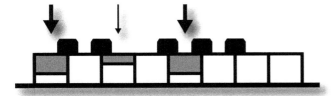

Status / Ch #	Note #	Pressure Value
(0 - 15)	(0 - 127)	(0 - 127)
(1101 CCCC)	(NNNN NNNN)	(VVVV VVVV)

assigned to such parameters as vibrato, loudness, filter cutoff and pitch. Two types of Pressure messages are defined by the MIDI spec:

- Channel Pressure
- Polyphonic Key Pressure

FIGURE 2.11
Channel Pressure messages simultaneously affect all notes that are being trans-mitted over a MIDI channel.

CHANNEL PRESSURE MESSAGES

Channel Pressure messages are commonly transmitted by instruments that respond only to a single, overall pressure, regardless of the number of keys that are being played at any one time (Figure 2.14). For example, if six notes are played on a keyboard and additional AfterTouch pressure is applied to just one key, the assigned parameter would be applied to all six notes. A Channel Pressure message consists of 3 bytes of information (Figure 2.11):

- Channel Pressure status/MIDI channel number
- MIDI note number
- Pressure value

POLYPHONIC KEY PRESSURE MESSAGES

Polyphonic Key Pressure messages respond to pressure changes that are applied to the individual keys of a keyboard. That's to say that a suitably equipped instru-ment can transmit or respond to individual Pressure messages for each key that's depressed (Figure 2.12). How a device responds to these messages will often vary from manufacturer to manufacturer (or can be assigned by the user); how-ever, pressure values are commonly assigned to such performance parameters as vibrato, loudness, timbre and pitch. Although controllers that are capable of producing polyphonic pressure are generally more expensive, it's not uncom-mon for an instrument to respond to these messages.

FIGURE 2.12
Byte structure of a Polyphonic Key Pressure message.

Status / Ch #	Note #	Pressure Value
(0 - 15)	(0 - 127)	(0 - 127)
(1010 CCCC)	(NNNN NNNN)	(VVVV VVVV)

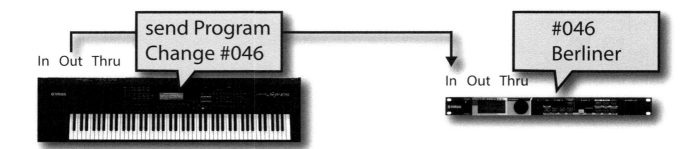

FIGURE 2.13
Program Change messages can be used to change sound patches remotely or automatically from a sequencer or a controller.

- Polyphonic Key Pressure status/MIDI channel number
- MIDI note number
- Pressure value

Program Change Messages

Program Change messages are used to change the active program or preset number of a MIDI instrument or device. A preset is a user- or factory-defined number that actively selects a specific sound patch or system setup. Using this extremely handy message, up to 128 presets can be remotely selected from another device or controller; for example:

- A Program Change message can be transmitted from a remote keyboard or controller to an instrument, allowing sound patches to be remotely switched (Figures 2.13 and 2.14).
- Program Change messages could be programmed at the beginning of a sequence to instruct the various instruments or voice generators to set to the correct sound patch before playing.
- A Program Change message could be used to alter patches on an effects device, either in the studio or on stage.
- A Program Change message (Figure 2.15) consists of 2 bytes of message information:
  - Program Change status/MIDI channel number (1–16)
  - Program ID number (0–127)

FIGURE 2.14
Workstations and sequencer software systems will often allow patches to be recalled via Program Change messages. (Courtesy of Steinberg Media Technologies GmbH, A Division of Yamaha Corporation, www.steinberg.net.)

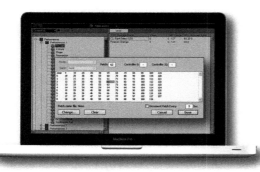

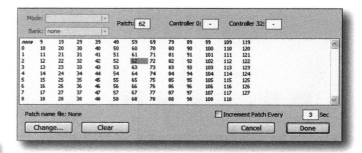

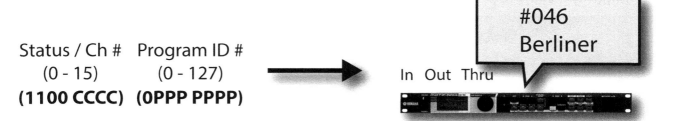

Status / Ch #   Program ID #
(0 - 15)   (0 - 127)
(1100 CCCC)   (0PPP PPPP)

In Out Thru

#046
Berliner

FIGURE 2.15
Byte structure of a Program
Change message.

Pitch Bend Messages

Pitch bend sensitivity refers to the response sensitivity (in semitones) of a pitch-bend wheel or other pitch-bend controller (which, as you'd expect, is used to bend the pitch of a note upward or downward). Since the ear can be extremely sensitive to changes in pitch, this control parameter is encoded using 2 data bytes, yielding a total of 16,384 steps. Since this parameter is most commonly affected by varying a pitch wheel (Figures 2.16 and 2.17), the control values range from –8192 to +8191, with 0 being the instrument's or part's unaltered pitch. Although the General MIDI spec recommends that the pitch-bend wheel have a range of ±2 half steps, most devices allow the pitch-bend range to be varied upwards to a full octave.

Control Change Messages

Control Change messages are used to transmit information to a device (either internally or through a MIDI line/network) that relates to real-time control over its performance parameters.

There are three types of Control Change messages that can be transmitted via MIDI:

- Continuous controllers – controllers that relay a full range of variable control settings (often ranging in value from 0 to 127, although, in certain cases, two controller messages can be combined in tandem to achieve a greater resolution).
- Switch controllers – controllers that have either an "off" or an "on" state with no intermediate settings.
- Channel mode message controllers – controllers that range from controller numbers 120 through 127 and are used to set the note sounding status, instrument reset, local control on/off, all notes off and MIDI mode status of a device or instrument.

FIGURE 2.16
Byte structure of a Pitch
Bend message.

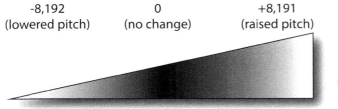

-8,192   0   +8,191
(lowered pitch) (no change)   (raised pitch)

Status / Ch #   Pitch Bend LSB   Pitch Bend MSB

(1111 CCCC)   (LLLL LLLL)   (MMMM MMMM)

Minimum Value = 0   Mid Value = 64   Maximum Value = 127

FIGURE 2.17
Pitch-bend wheel data value ranges.

A single Control Change message or a stream of such messages is transmitted whenever controllers (such as foot switches, foot pedals, pitch-bend wheels, modulation wheels, or breath controllers) are varied in real time. Newer controllers and software editors often offer up a wide range of switched and variable controllers, allowing for extensive, user-programmable control over any number of device, voice and mixing parameters in real time (Figure 2.18).

A Control Change message (Figure 2.19) consists of 3 bytes of information:

- Control Change status/MIDI channel number (1–16)
- Controller ID number (0–127)
- Corresponding controller value (0–127)

CONTROLLER ID NUMBERS

FIGURE 2.18
Komplete Kontrol S49 MIDI controller. (Courtesy of Native Instruments GMBH, www.native-instruments.com.)

As you can see, the second byte of the Control Change message is used to denote the controller ID number. This all-important value is used to specify which of the device's program or performance parameters are to be addressed. Before we move on to discuss the various control numbers that are listed in the MIDI spec, it's best to take time out to discuss the values that are assigned to these parameters (i.e., the third and subsequent bytes).

controllers

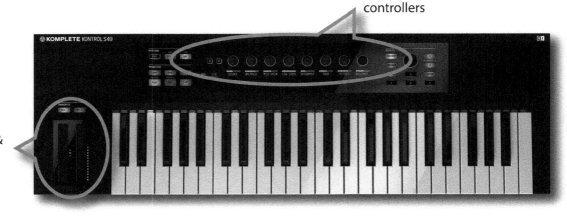

pitch bend & modulation

Status / Ch #	Controller ID #	Controller Value
(0 - 15)	(0 - 127)	(0 - 127)
(1011 CCCC)	(0CCC CCCC)	(0VVV VVVV)

CONTROLLER VALUES

The third byte of the Control Change message is used to denote the controller's actual data value. This value is used to specify the position, depth or level of a parameter. Here are a few examples as to how these values can be implemented to vary, control and mix parameters using several in-screen and hands-on hardware controls (Figure 2.20):

FIGURE 2.19
Byte structure of a Control Change message.

- In the case of a variable control parameter that doesn't require that the settings be made in extremely fine increments, a 7-bit continuous controller allows for values over a 128-step range (with 0 being the minimum and 127 being the maximum value).
- The value range of the continuous controller falls between 0 and 127, with a value of 64 representing a mid or center position.
- The value range of a switch controller is often 0 (off) and 127 (on); however, it's not uncommon for a switching function to respond to continuous controller messages by recognizing the value range of 0 to 63 as being "off" and 64 to 127 as being "on".

In certain cases, a single, 7-bit "course" message (128 steps) might not be enough to adequately manage a controller's resolution. For this reason, the MIDI spec allows the resolution of a number of parameters to be increased by adding an additional "fine" controller value message to the data stream, resulting in a resolution that yields an overall total of 16,384 discrete steps!

FIGURE 2.20
Continuous controller messages can be varied in real-time or under automation using a number of input methods.

on-screen controls

hands-on hardware controls

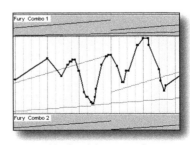

rubberband draw controls

EXPLANATION OF CONTROLLER ID PARAMETERS

The following sections detail the general categories and conventions for assigning controller numbers to an associated parameter, as specified by the 1995 update of the MIDI Manufacturers Association (MMA, www.midi.org). An overview of these controllers can be seen in Table 2.2. This is definitely an important chart to earmark, as these numbers will be an important guide to knowing or finding the right ID number that can help you on your path toward finding that perfect variable to make it sound right.

Bank Select

Nowadays, it's common for an instrument or device to have more than 128 presets, patches and general settings. Since MIDI Program Change messages can only switch between 128 settings, the Bank Select controller (sometimes called Bank Switch) can be used to switch between multiple groupings of 128 presets. As an example, a device that has 256 presets could be easily divided into 2 banks of 128. Thus, if you wanted to call up preset 129, you would transmit a Bank Select message that would switch the system to its second bank and then follow this with a Program Change message that would select the first program in that bank. This parameter can be used on different channels to affect each part within a multi-timbral instrument.

MOD Wheel
Numbers: 1 (coarse), 33 (fine)

This message is used to set the modulation (or MOD) wheel value. When modulation is used, a vibrato effect is introduced, often by introducing a low-frequency oscillator (LFO) effect. Although a fine modulation message can be used to give additional control, many units will often ignore the presence of this message. This parameter can be used on different channels to affect each part within a multi-timbral instrument.

Breath Control
Numbers: 2 (coarse), 34 (fine)

Breath Control is often used by wind players to add expression to a performance. As such, this message can be user assigned to control a wide range of parameters (such as AfterTouch, modulation or expression). This parameter can be used on different channels to affect each part within a multi-timbral instrument.

Foot Pedal
Numbers: 4 (coarse), 36 (fine)

This controller often makes use of a foot pedal, which uses a potentiometer to continuously control such user-assignable parameters as AfterTouch, modulation and expression. Although a foot controller can vary in value between 0 and 127, if it's being used as a foot switch, values ranging from 0 to 63 will be

Table 2.2	Controller ID Numbers, Outlining Both the Defined Format and Conventional Controller Assignments	

Control Number	Parameter	
14-Bit Controllers Coarse	**Most Significant Bit (MSB)**	
0	Bank Select	0–127
1	Modulation Wheel or Lever	0–127
2	Breath Controller	0–127
3	Undefined	0–127
4	Foot Controller	0–127
5	Portamento Time	0–127
6	Data Entry MSB	0–127
7	Channel Volume (formerly Main Volume)	0–127
8	Balance	0–127
9	Undefined	0–127
10	Pan	0–127
11	Expression Controller	0–127
12	Effect Control 1	0–127
13	Effect Control 2	0–127
14	Undefined	0–127
15	Undefined	0–127
16–19	General Purpose Controllers 1–4	0–127
20–31	Undefined	0–127
14-Bit Controllers Fine	**Least Significant Bit (LSB)**	
32	LSB for Control 0 (Bank Select)	0–127
33	LSB for Control 1 (Modulation Wheel or Lever)	0–127
34	LSB for Control 2 (Breath Controller)	0–127
35	LSB for Control 3 (Undefined)	0–127
36	LSB for Control 4 (Foot Controller)	0–127
37	LSB for Control 5 (Portamento Time)	0–127
38	LSB for Control 6 (Data Entry)	0–127
39	LSB for Control 7 (Channel Volume, formerly Main Volume)	0–127
40	LSB for Control 8 (Balance)	0–127
41	LSB for Control 9 (Undefined)	0–127

Control Number	Parameter
14-Bit Controllers Coarse	**Most Significant Bit (MSB)**

Table 2.2 Continued

Control Number		Parameter
	14-Bit Controllers Coarse	**Most Significant Bit (MSB)**
42	LSB for Control 10 (Pan)	0–127
43	LSB for Control 11 (Expression Controller)	0–127
44	LSB for Control 12 (Effect Control 1)	0–127
45	LSB for Control 13 (Effect Control 2)	0–127
46–47	LSB for Control 14–15 (Undefined)	0–127
48–51	LSB for Control 16-19 (General Purpose Controllers 1–4)	0–127
52–63	LSB for Control 20–31 (Undefined)	0–127
	7-Bit Controllers	
64	Damper Pedal On/Off (Sustain)	<63 off, >64 on
65	Portamento On/Off	<63 off, >64 on
66	Sostenuto On/Off	<63 off, >64 on
67	Soft Pedal On/Off	<63 off, >64 on
68	Legato Footswitch	<63 normal, >64 legato
69	Hold 2	<63 off, >64 on
70	Sound Controller 1 (default: Sound Variation)	0–127 LSB
71	Sound Controller 2 (default: Timbre/ Harmonic Intens.)	0–127 LSB
72	Sound Controller 3 (default: Release Time)	0–127 LSB
73	Sound Controller 4 (default: Attack Time)	0–127 LSB
74	Sound Controller 5 (default: Brightness)	0–127 LSB
75	Sound Controller 6 (default: Decay Time; see MMA RP-021)	0–127 LSB
76	Sound Controller 7 (default: Vibrato Rate; see MMA RP-021)	0–127 LSB
77	Sound Controller 8 (default: Vibrato Depth; see MMA RP-021)	0–127 LSB
78	Sound Controller 9 (default: Vibrato Delay; see MMA RP-021)	0–127 LSB
79	Sound Controller 10 (default undefined; see MMA RP-021)	0–127 LSB
80–83	General Purpose Controller 5–8	0–127 LSB

Control Number	Parameter
14-Bit Controllers Coarse	**Most Significant Bit (MSB)**
84	Portamento Control	0–127 LSB
85–90	Undefined	–
91	Effects 1 Depth (default: Reverb Send Level)	0–127 LSB
92	Effects 2 Depth (default: Tremolo Level)	0–127 LSB
93	Effects 3 Depth (default: Chorus Send Level)	0–127 LSB
94	Effects 4 Depth (default: Celeste \<Detune\> Depth)	0–127 LSB
95	Effects 5 Depth (default: Phaser Depth)	0–127 LSB
Controllers	
96	Data Increment (Data Entry +1)	–
97	Data Decrement (Data Entry –1)	–
98	Non-Registered Parameter Number (NRPN)	0–127 LSB
99	Non-Registered Parameter Number (NRPN)	0–127 MSB
100	Registered Parameter Number (RPN)	0–127 LSB
101	Registered Parameter Number (RPN)	0–127 MSB
102–119	Undefined	–
Reserved for Channel Mode Messages	
120	All Sound Off	0
121	Reset All Controllers	
122	Local Control On/Off	(0 off, 127 on)
123	All Notes Off	
124	Omni Mode Off	
125	Omni Mode On	
126	Poly Mode On/Off	
127	Poly Mode On	

Note: Bank Select can sometimes be used to switch between drum kits on multi-timbral devices that offer up percussion.

interpreted as being "off", while those ranging from 64 to 127 will be interpreted as being "on". This parameter can be used on different channels to affect each part within a multi-timbral instrument.

Portamento Time
Numbers: 5 (coarse), 37 (fine)

Portamento can be used to slide one note into another, rather than having the second note immediately follow the first. This value is used to determine the rate of slide (in time) between the two notes. This parameter can be used on different channels to affect each part within a multi-timbral instrument.

Data Entry
Numbers: 6 (coarse), 38 (fine)

This message value can be used to vary a Registered or Non-Registered Parameter. Registered Parameter Numbers (RPNs) and Non-Registered Parameter Numbers (NPRNs) are used to affect parameters that are specific to a particular MIDI device or aren't controllable by a defined Control Change message. For example, Pitch Bend Sensitivity and Fine Tuning parameters are RPN parameters. Varying these values can be used to remotely alter these settings.

Note: certain devices aren't able to control RPN or NRPN messages using this parameter. In such a case, it's often possible to directly assign a defined parameter to this message type.

Channel Volume (Formerly Known as Main Volume)
Numbers: 7 (coarse), 39 (fine)

This parameter affects the volume level of a device or individual musical parts (in the case of a multi-timbral instrument). This parameter is independent of individual note velocities and allows for the overall part or device volume that's transmitted over a channel to be altered.

Note: fine resolution messages (controller 39) are rarely supported by most instruments and mixing devices. If a non-supporting device receives these messages, they will simply be ignored.

Stereo Balance
Numbers: 8 (coarse), 40 (fine)

A Stereo Balance controller is used to vary the relative levels between two independent sound sources (Figure 2.21). As with the balance control on a stereo preamplifier, this controller is used to set the relative left/right balance of a stereo signal (unlike pan, which is used to place a mono signal within a stereo field). The value range of this controller falls between 0 (full left sound source) and 127 (full right sound source), with a value of 64 representing an equally balanced stereo field.

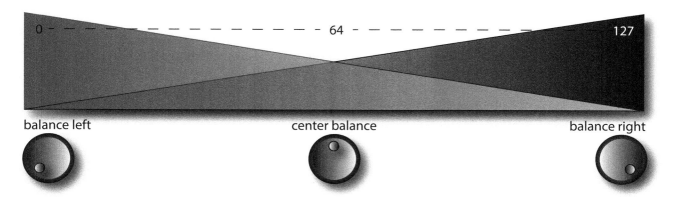

Pan
Numbers: 10 (coarse), 42 (fine)

A Pan controller is used to position the relative balance of a single sound source between the left and right channels of a stereo sound field (Figure 2.22). The value range of this controller falls between 0 (hard left) and 127 (hard right), with a value of 64 representing a balanced center position. Pan is a common parameter, in that it can be used to internally mix all of the parts of a multi-timbral or sound-mixing device to a single, bused output.

Expression
Numbers: 11 (coarse), 43 (fine)

An Expression controller is used to accent the existing level settings of a MIDI instrument or device (e.g., when a crescendo or decrescendo is needed). This control works by dividing a part's overall volume into 128 steps (or 16,384 if fine resolution is used), thereby making it possible to alter level over the course of a passage, without having to change the velocity levels of each note within that passage. Whenever the Expression control is set to its maximum value (i.e., 127 in coarse resolution), the output level will be set to the part's actual main volume setting. As the values decrease, the volume settings will proportionately decrease; that is, a value of 64 will reduce the signal to 50% of its full level, and 0 will reduce the output to 0% of its channel volume level (effectively turning the gain all the way down). It should be noted that the fine controller byte (43) is often ignored on most receiving systems.

FIGURE 2.21
Stereo Balance controller data value ranges and corresponding settings.

FIGURE 2.22
Pan controller data value ranges.

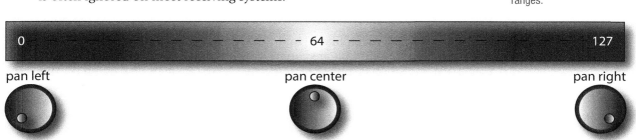

Effect Control 1 and 2
Numbers: 12 and 13 (coarse), 44 and 45 (fine)

These controllers can be used to vary any parameter relating to an effects device (such as the reverb decay time for a reverb unit that's built into a GM sound module). This parameter can be used on different channels to affect each part within a multi-timbral instrument.

Note: controllers 91 through 95 can be used to vary overall effect level, chorus and phaser.

General Purpose Controllers
Numbers: 16 through 19 (coarse), 48 and 51 (fine)

These controllers are open to use for varying any parameter that can be assigned to them.

Note: fine-resolution General Purpose messages are rarely supported by most instruments and mixing devices.

Hold Pedal
Number: 64

This control is used to sustain (hold) notes that are currently being played. When on, the notes will continue to play, even though the keys have been released, and Note-Off messages will be transmitted only after the hold pedal has been switched off. This parameter can be used on different channels to affect each part within a multi-timbral instrument.

The remaining controller parameters are capable only of being encoded using a single, 7-bit "coarse" message that will yield a total of 128 continuous steps or can be translated in value ranges that represent a switched on/off state.

Portamento On/Off
Number: 65

This control determines whether the portamento effect is on or off. This parameter can be used on different channels to affect each part within a multi-timbral instrument.

Sostenuto Pedal
Number: 66

This control works much like a Hold Pedal message, with the exception that it will only sustain notes that are currently being played (i.e., Note-On messages have been transmitted, but their respective Note-Off messages haven't). This has the effect of holding keys that were initiated when the pedal was pressed while not holding keys that are subsequently played during the sustain time. For example, this pedal could be used to hold an initial chord while not holding a melody that's being played around it. This parameter can be used on different channels to affect each part within a multi-timbral instrument.

Soft Pedal
Number: 67

When on, the volume of any played note is lowered. This parameter can be used on different channels to affect each part within a multi-timbral instrument.

Legato Pedal
Number: 68

This control is used to create a legato (or smooth transition) effect between notes. This is often achieved by not sounding the attack portion of the part's voltage-controlled amplifier (VCA) envelope. This effect can be used to simulate the phrasing of wind and brass players, as well as guitar pull-offs and hammer-ons (where secondary notes are not picked). This parameter can be used on different channels to affect each part within a multi-timbral instrument.

Hold 2 Pedal
Number: 69

Unlike the other Hold Pedal controller, this control doesn't permanently sustain a sounded note until the musician releases the pedal. It is used to lengthen the release time of a played note, effectively increasing the note's fade-out time by lengthening the VCA's release time.

Sound Variation (Sound Controller 1)
Number: 70

This control affects any parameter that's associated with the reproduction of sound. Sound Controller 1 can be used to alter or tune a sound file during playback by varying its sample rate. This parameter can be used on different channels to affect each part within a multi-timbral instrument.

Sound Timbre (Sound Controller 2)
Number: 71

This control affects any parameter that's associated with the reproduction of sound. Sound Controller 2 can be used as a brightness control by varying the envelope of a voltage-controlled filter (VCF).

Note: VCA and VCF parameters can also be varied through the use of other controllers.

Sound Release Time (Sound Controller 3)
Number: 72

This controls the length of time that it takes a sound to fade out by altering the VCA's release time envelope. This parameter can be used on different channels to affect each part within a multi-timbral instrument.

Note: VCA and VCF parameters can also be varied through the use of other controllers.

Sound Attack Time (Sound Controller 4)
Number: 73

This controls the length of time that it takes a sound to fade in, by altering the VCA's attack time. This parameter can be used on different channels to affect each part within a multi-timbral instrument.

Note: VCA and VCF parameters can also be varied through the use of other controllers.

Sound Brightness (Sound Controller 5)
Number: 74

This parameter acts as a timbral brightness control by varying the VCF's cutoff filter frequency. This parameter can be used on different channels to affect each part within a multi-timbral instrument.

Note: VCA and VCF parameters can also be varied through the use of other controllers.

Sound Controller 6, 7, 8, 9 and 10
Numbers: 75, 76, 77, 78 and 79

These five additional controllers can be freely used to affect any assignable parameter that's associated with the audio production circuitry of a sound module. This parameter can be used on different channels to affect each part within a multi-timbral instrument.

General Purpose Buttons
Numbers: 80, 81, 82 and 83

These four General Purpose buttons can be freely used to affect any assignable on/off parameter within a device. This parameter can be used on different channels to affect each part within a multi-timbral instrument.

Portamento Control
Number: 84

A Portamento Control message (in conjunction with a MIDI note number) is used to indicate the starting note of a portamento (whereby one note slides into the next rather than having the second note immediately follow the first). The slide rate will be set by the Portamento Time (Controller 5), and the current status of the Portamento On/Off (Controller 65) is ignored.

Effects Level
Number: 91

This control is used to control the effect level or wet/dry balance for an instrument or part (often referring to its reverb or delay level). For devices with built-in effects, this parameter can be used on different channels to affect each part within a multi-timbral instrument.

Tremolo Level
Number: 92

This control is used to control the tremolo level within an instrument or part. This parameter can be used on different channels to affect each part within a multi-timbral instrument.

Chorus Level
Number: 93

By combining two identical (and often slightly delayed) signals that are slightly detuned in pitch from one another, another effect known as chorusing can be created. Chorusing is an effects tool that's often used by guitarists, vocalists and other musicians to add depth, richness and harmonic structure to their sound. This control is used to control the chorus level or wet/dry balance for an instrument or part. This parameter can be used on different channels to affect each part within a multi-timbral instrument.

Celeste Level
Number: 94

This control is used to control the celeste (detune) level for an instrument or part. This parameter can be used on different channels to affect each part within a multi-timbral instrument.

Phaser Level
Number: 95

This control is used to control a phasing effect level or wet/dry balance for an instrument or part. This parameter can be used on different channels to affect each part within a multi-timbral instrument.

Data Button Increment
Number: 96

This control causes a parameter (most often a Registered or Non-Registered Parameter) to increase its current value by 1.

Data Button Decrement
Number: 97

This control causes a parameter (most often a Registered or Non-Registered Parameter) to decrease its current value by one.

Non-Registered Parameter Number (NRPN)
Numbers: 99 (coarse), 98 (fine)

This control determines which Data Button or Data Entry controller will be incremented or decremented. The assignment of this parameter is entirely up to the manufacturer and doesn't have to be registered with the International MIDI Association (IMA).

Registered Parameter Number (RPN)
Numbers: 101 (coarse), 100 (fine)

This control determines which Data Button or Data Entry controller will be incremented or decremented. The functional parameter assignments and their designations are determined by the IMA.

All Sound Off
Number: 120

This control is used to turn off all sounding notes that were turned on by received Note-On messages but which haven't yet been turned off by respective Note-Off messages.

All Controllers Off
Number: 121

This control is used to reinitialize all of the controllers (continuous, switch and incremental) within one or more receiving MIDI instruments or devices to their standard, power-up default state. Upon receipt of this message, all switched parameters will be turned off and continuous controllers will be set to their minimum values. This parameter can be used on different channels to affect each part within a multi-timbral instrument.

Local Keyboard On/Off
Number: 122

FIGURE 2.23
The Local Keyboard On/Off function is used to disconnect a device's performance controller from its internal sound generators.

The Local Keyboard (Local Control) On/Off message is used to disconnect the controller of a MIDI instrument from its own internal voice generators (Figure 2.23). This feature is useful for turning a keyboard instrument into a master controller by disconnecting its keyboard from its sound-generating

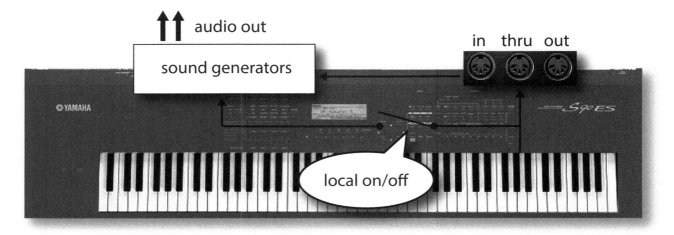

circuitry. For example, when the instrument's local control is switched off, a synth can be used to output MIDI to a sequencer or other devices without having to listen to its own internal sounds.

With the Local turned off, you could easily route MIDI through your computer/sequencer and back into the device's own internal voice generators. In short, the local control feature splits your instrument into two parts: a master controller for playing other instruments in the system and a performance instrument that can be viewed as any other instrument in the MIDI setup. It's a setup that effectively makes the system more flexible while reducing setup and performance conflicts.

It should be noted that turning the Local switch to "off" is important within a sequencer/DAW production setup, as this allows the respective devices, instruments and parts within a setup to be heard without listening to sounds that would otherwise be playing from the master synth. Of course, if you're using a MIDI controller this setting doesn't apply, as it has no internal sound generators. A Local Control message consists of 2 bytes of information: a MIDI channel number and a Local Control on/off status byte.

All Notes Off
Number: 123

Occasionally, a MIDI instrument will receive a Note-On message, and (by some technical glitch) the following Note-Off message is somehow ignored or not received. This unfortunate event often results in a stuck note that continues to sound until a Note-Off message is received for that note. As an alternative to frantically searching for the right Note-Off key on the right MIDI channel, an All Notes Off "panic message" can be transmitted that effectively turns off all 128 notes on all channels and ports. Often, a MIDI interface or sequencer will include a button that can globally transmit this message throughout the connected system.

Omni Mode Off
Number: 124

Omni Mode Off refers to how an instrument will respond to MIDI messages at its input. Upon the reception of an Omni Mode Off message, a MIDI instrument or device will only respond to a single MIDI channel or set of assigned channels (in the case of a multi-timbral instrument).

Omni Mode On
Number: 125

Omni Mode On refers to how an instrument will respond to MIDI messages at its input. Upon the reception of an Omni Mode On message, a MIDI instrument or device will respond to all channel messages that are being received, regardless of its MIDI channel assignment.

Monophonic Operation
Number: 126

Upon receiving a Mono Mode On message, a MIDI instrument will assign individual voices to the consecutive MIDI channels, starting from the lowest currently assigned or base channel. The instrument will then be limited to playing only one note at a time on each MIDI channel, even if it's capable of playing multiple notes at any one time.

Polyphonic Operation
Number: 127

Upon receiving a Poly Mode On message, a MIDI instrument will assign voices to the consecutive MIDI channels, starting from the lowest currently assigned or base channel. The instrument will then respond to MIDI channels polyphonically, allowing the device to play more than one note at a time over a given channel or number of channels.

System Messages

As the name implies, System messages are globally transmitted to every MIDI device in the MIDI chain. This is accomplished because MIDI channel numbers aren't addressed within the byte structure of a System message. Thus, any device will respond to these messages, regardless of its MIDI channel assignment. The three System message types are System-Common messages, System Real-Time messages and System-Exclusive messages.

System-Common Messages

System-Common messages are used to transmit MIDI time code, song position pointer, song select, tune request and end-of-exclusive data messages throughout the MIDI system or 16 channels of a specified MIDI port.

MTC QUARTER-FRAME MESSAGES

MIDI time code (MTC) provides a cost-effective and easily implemented way to translate SMPTE (a standardized synchronization time code) into an equivalent code that conforms to the MIDI 1.0 spec. It allows time-based codes and commands to be distributed throughout the MIDI chain in a cheap, stable and easy-to-implement way. MTC Quarter-Frame messages are transmitted and recognized by MIDI devices that can understand and execute MTC commands. A grouping of eight quarter frames is used to denote a complete time-code address (in hours, minutes, seconds and frames), allowing the SMPTE address to be updated every two frames. Each Quarter-Frame message contains 2 bytes. The first is a quarter-frame common header, while the second byte contains a 4-bit nibble that represents the message number (0–7). A final nibble is used to encode the time field (in hours, minutes, seconds or frames). More in-depth coverage of MIDI time code can be found in Chapter 10.

SONG POSITION POINTER MESSAGES

As with MIDI time code, the song position pointer (SPP) lets you synchronize a sequencer, tape recorder or drum machine to an external source from any measure position within a song. The SPP message is used to reference a location point in a MIDI sequence (in measures) to a matching location within an external device. This message provides a timing reference that increments once for every six MIDI clock messages (with respect to the beginning of a composition). Unlike MTC (which provides the system with a universal address location point), SPP's timing reference can change with tempo variations, often requiring that a special tempo map be calculated in order to maintain synchronization. Because of this fact, SPP is used far less often than MIDI time code. SPP messages are generally transmitted while the MIDI sequence is stopped, allowing MIDI devices equipped with SPP to chase (in a fast-forward motion) through the song and lock to the external source once relative sync is achieved. More in-depth coverage of the SPP can be found in Chapter 11.

SONG SELECT MESSAGES

Song Select messages are used to request a specific song from a drum machine or sequencer (as identified by its song ID number). Once selected, the song will thereafter respond to MIDI Start, Stop and Continue messages.

TUNE REQUEST MESSAGES

The Tune Request message is used to request that a MIDI instrument initiate its internal tuning routine (if so equipped).

END OF EXCLUSIVE MESSAGES

The transmission of an End of Exclusive (EOX) message is used to indicate the end of a System-Exclusive message. In-depth coverage of System-Exclusive messages will be discussed later in this chapter and in Chapter 4.

System Real-Time Messages

Single-byte System Real-Time messages provide the precise timing element required to synchronize all of the MIDI devices in a connected system. To avoid timing delays, the MIDI specification allows System Real-Time messages to be inserted at any point in the data stream, even between other MIDI messages (Figure 2.24).

TIMING CLOCK MESSAGES

The MIDI Timing Clock message is transmitted within the MIDI data stream at various resolution rates. It is used to synchronize the internal timing clocks of each MIDI device within the system and is transmitted in both the Start and Stop modes at the currently defined tempo rate. In the early days of MIDI, these

note on
note number
velocity
note on
note number
velocity
......
timing clock
note on
velocity
note off
note on

In Out Thru   In Out Thru

FIGURE 2.24
System Real-Time messages can be inserted within the byte stream of other MIDI messages.

rates (which are measured in pulses per quarter note [ppq]) ranged from 24 to 128 ppq; however, continued advances in technology have brought these rates up to 240, 480 or even 960 ppq.

Start Messages

Upon receipt of a timing clock message, the MIDI Start command instructs all connected MIDI devices to begin playing from their internal sequences initial start point. Should a program be in mid-sequence, the Start command will reposition the sequence back to its beginning, at which point it will begin to play.

Stop Messages

Upon receipt of a MIDI Stop command, all devices within the system will stop playing at their current position point.

Continue Messages

After receiving a MIDI Stop command, a MIDI Continue message will instruct all connected devices to resume playing their internal sequences from the precise point at which it was stopped.

Active Sensing Messages

When in the Stop mode, an optional Active Sensing message can be transmitted throughout the MIDI data stream every 300 milliseconds. This instructs devices that can recognize this message that they're still connected to an active MIDI data stream.

System Reset Messages

A System Reset message is manually transmitted in order to reset a MIDI device or instrument back to its initial power-up default settings (commonly mode 1, local control on and all notes off).

System-Exclusive Messages

The System-Exclusive (SysEx) message lets MIDI manufacturers, programmers and designers communicate customized MIDI messages between MIDI devices.

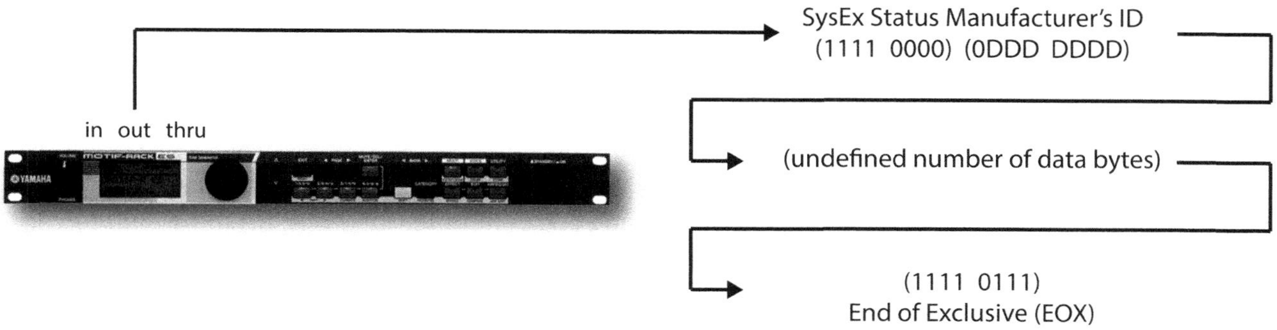

SysEx Status Manufacturer's ID
(1111 0000) (0DDD DDDD)

in out thru

(undefined number of data bytes)

(1111 0111)
End of Exclusive (EOX)

These messages give manufacturers, programmers and designers the freedom to communicate any device-specific data of an unrestricted length as they see fit. SysEx data is commonly used for the bulk transmission and reception of program/patch data, sample data and real-time control over a device's parameters. The transmission format of a SysEx message (Figure 2.25) as defined by the MIDI standard includes a SysEx status header, manufacturer's ID number, any number of SysEx data bytes and an EOX byte. Upon receiving a SysEx message, the identification number is read by a MIDI device to determine whether or not the following messages are relevant. This is easily accomplished, because a unique 1- or 3-byte ID number is assigned to each registered MIDI manufacturer. If this number doesn't match the receiving MIDI device, the ensuing data bytes will be ignored. Once a valid stream of SysEx data is transmitted, a final EOX message is sent, after which the device will again begin responding to incoming MIDI performance messages. A detailed practical explanation of the many uses (and wonders) of SysEx can be found in Chapter 9. I definitely recommend that you check these out.

FIGURE 2.25
System-exclusive data.

UNIVERSAL NON-REAL-TIME SYSTEM EXCLUSIVE

Universal Non-Real-Time SysEx data is a protocol that's used to communicate control and non-real-time performance data. It's currently used to intelligently communicate a data-handshaking protocol that informs a device that a specific event is about to occur or that specific data is about to be requested. It is also used to transmit and receive universal sample-dump data or to transmit MIDI time-code cueing messages. A universal Non-Real-Time SysEx message consists of 4 or 5 bytes that include 2 sub-ID data bytes that identify which non-real-time parameter is to be addressed. It is then followed by a stream of pertinent SysEx data.

UNIVERSAL REAL-TIME SYSTEM EXCLUSIVE

Currently, two Universal Real-Time SysEx messages are defined. Both of them relate to the MTC synchronization code (which is discussed in detail in Chapter 11). These include the full message data and user-bit data found in a SMPTE address.

Running Status

Within the MIDI 1.0 specification, special provisions have been made to reduce the need for conveying redundant MIDI data. This mode, known as *Running Status*, allows a series of consecutive MIDI messages that have the same status byte type to be communicated without repeating the same status byte each time a MIDI message is sent. For example, we know that a standard MIDI message is made up of both a status byte and one or more data bytes. When using running status, a series of Pitch Bend messages that have been generated by a controller would transmit an initial status and data byte message, followed only by a series of related data (pitch-bend level) bytes, without the need for including redundant status bytes. The same could be said for Note-On, Note-Off or any other status message type. Although the transmission of Running Status messages is optional, all MIDI devices must be able to identify and respond to this data transmission mode.

CHAPTER 3

The MIDI 2.0 Spec

The fact that MIDI 1.0 has been in place from 1983 till 2020 (almost 40 years!) is an amazing testament to the individuals and companies that joined forces to create a robust spec that has forever changed the face of music production. However, as with all things, the time has finally come to broaden the spec, so as to more fully take advantage of the advances that have taken place in music hardware, music software, DAW capabilities and overall data distribution.

At the time of this writing, MIDI 2.0 has just been formally ratified and released by the MIDI Manufacturers Association (MMA) at the Winter NAMM, 2020, and was also finalized by the Association of Musical Electronics Industry (AMEI) of Japan. It was pointed out at this show by a good friend of mine with the Yamaha Corporation of America that the specifics of MIDI 2.0 could easily fill four books by itself. For this reason, I feel it best that I present you with an overall summary of the capabilities, strengths and technology of the 2.0 spec … and not the full details of the spec itself. Of course, those of you who have a need to dig deeper into the spec can do so by logging onto the MMA website (www.midi.org) and downloading the current MIDI 2.0 spec or other related MIDI subsets as they become available.

INTRODUCTION TO MIDI 2.0

The result of a global, decade-long development, MIDI 2.0 is an effort to keep MIDI relevant in the future, using a new Universal MIDI Packet format that allows for easier communication over any digital medium (such as USB, Thunderbolt, Ethernet or any other medium that's yet to be envisioned). Future growth and flexibility is also a big part of 2.0, in that additional space has also been reserved for other message and controller types that might be needed in the future.

Some of the more important additions to the MIDI 2.0 spec include:

- Bi-directional communication
- Backwards compatibility

- Both protocol specs are fully supported
- Higher resolution for velocity and control messages
- Tighter timing
- Sixteen channels become 256
- Built-in support for "per-note events"

THE THREE BS

To begin with, let's take a look at the basic foundation of the MIDI 2.0 spec, known as "The Three Bs":

- Bi-directional communication
- Backwards compatibility
- Both protocols (simultaneous support for both specs)

Bi-Directional Communication

As we've read in Chapter 2, MIDI 1.0 is only capable of being sent in one direction. It is only able to communicate from a source to a receiving destination, without the ability to know what the capabilities of the destination device are or if that information can be properly interpreted. By contrast, MIDI 2.0 can both communicate bi-directionally (Figure 3.1), as well as set up a dialog that allows the involved devices to be automatically configured in ways that allow them to work together in better and/or faster ways.

Some of you might remember that in the past, certain samplers and SCSI-capable devices made use of tiny jumper switches that had to be set with a screwdriver or a pen point, so as to make sure that the receiver could properly receive the data that was being sent. Now, with 2.0, these compatibility settings can be automatically made with a digital handshake within the data stream itself. Just plug the devices in and when the connection is made, the device capabilities are automatically negotiated and configured, so as to best make use of their capabilities.

With 2.0's bi-directional capabilities, a device could ask "Hey, who are you and what are you capable of?" … when it hears back, they can then agree to talk to each other in a way that makes communications faster, better and easier for the task at hand.

FIGURE 3.1
MIDI 2.0 communicates data in a bi-directional manner between compatible devices.

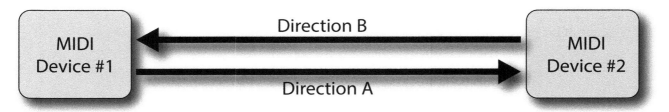

Backwards Compatibility

In the above scenario, if a 2.0 capable device shouts out to another device asking "hey, who are you?" and hears nothing back, it'll assume that the destination device is older and will automatically revert to using the fully supported language of MIDI 1.0.

Since the overwhelming majority of electronic instruments, controllers and processors are built upon the 1.0 spec, it stands to reason that communication with the entire, present-day music-making ecosystem must be fully and transparently supported.

Both Protocols

One of the core goals of MIDI 2.0 is to enhance the feature set of MIDI 1.0 whenever possible.

Higher Resolution for Velocity and Control Messages

One of the biggest developments in 2.0 is its expansion from 7-bit values (128 discrete steps) to 16-bit (which in the case of velocity represents a jump up to 65,536 discrete encoding steps). The list of available controllers has been expanded to over 32,000 different controller options, while their values have been enhanced by offering a 32-bit range (representing 4,294,967,296 possible discrete steps). This can be thought of as going from having rough control over every aspect of MIDI 1.0 to having extremely precise control over such articulation and control aspects, such as volume, modulation, pitch-bend, after-touch, etc. Potentially, this increased degree of accuracy will allow for greater control over performance values, making the process of music production more analog and fine-tunable in nature.

Since MIDI's inception, guitar, violin, woodwind and brass players often had to learn how to best express their instrument using the keyboard playing surface. Now, it should be possible to design newer, more sensitive controllers or translation pickups that would allow players to use their own instruments (or ones that are closer in form and playability to their own) in order to capture their sound into their favorite DAW, processing and/or printing software.

It could be argued that these higher degrees of resolution might be unnecessary for making music. However, newer types of instruments that are able to make use of micro-tonal performance characteristics to better mimic human performances would certainly benefit from the added control. In addition, MIDI is able to capture and control more than just musical data; its use in lighting and animatronics, for example, could certainly benefit from the added degree of resolution. In short, who knows what will be invented in the future that can best make use of the added capabilities and resolutions that MIDI 2.0 has to offer … time will tell.

Tighter Timing

MIDI 2.0 has improved timing accuracy for making sure that notes and other event types occur at precisely the right time. Through the use of Jitter Reduction

(JR) Timestamps, the precise time of an event can be directly encoded within the MIDI message itself. In this way, if a message gets delayed within the data stream, the system would be able to accurately keep track of the time that an event is supposed to take place.

Timing problems can often crop up when a great deal of MIDI data is passed through a single connection (data clogging). One situation where this might occur is when a great deal of data passes from a MIDI guitar to a sequenced track or synth instrument. In such a case, data might get clogged and latency delays might cause sounds to stutter or be mistimed. By time-stamping the actual MIDI data, it would actually be possible for the data to be returned to its proper position with a sequence … after it's been recorded.

Sixteen Channels Become 256

The original 1.0 spec offers up 16 separate channels over a single MIDI cable or data line. Each message sent over the 2.0 spec includes a channel group (of which there are 16) which can address any of 16 channels, allowing for the encoding of up to 256 distinct channels over a single cable or data line!

MIDI 1.0 data messages, such as System Exclusive messages, can also be transmitted along any one of these 256 channels, alongside MIDI 2.0 data, allowing for full compatibility between the newer and older specifications.

Built-in Support for "Per-Note" Events

Another significant advancement within 2.0 is the addition of new Per-Note messages which expand on the concepts adopted by the MIDI Manufacturers Association in MPE (MIDI Polyphonic Expression). This means that 2.0 will directly encode performance data from specialized instruments that are designed to offer extended controller functions over pitch-bend, expression, modulation, etc. all on a per-note basis. For example, newer-specialized keyboards and performance controllers (Figure 3.2) allow the artist to add pitch and other controller gestures to a single keyboard note or controller pad in a fully polyphonic fashion. In the future, the added capabilities of MIDI 2.0 over the encoding of controller data could take these and other performance/control functions even further.

Before the adoption of 2.0, the MPE-encoded messages were sent by rotating through a defined contiguous block of channels called Per-Note channels. These Per-Note messages were limited to Note-On, Note-Off, Channel Pressure (finger pressure), Pitch Bend (X-axis movement) and Controller #74 (Y-axis movement). All other messages (such as Program Change, volume and sustain) apply to all voices and are sent over a separate "Common" channel (which is usually CH 1 or 16). If channel 1 is used as the common channel, the Per-Note channels are 2–16. If 16 is used, the Per-Note channels will be 1–15. As an example,

FIGURE 3.2
ROLI Seaboard Block Studio Edition 24-note USB/Bluetooth LE Keyboard Controller with MPE. Polyphonic Aftertouch/Pitch Bend and ROLI Studio Software. (Courtesy of ROLI, www. roli.com), photo © ROLI Ltd. 2020.

it's also possible, using MIDI 1.0, to split an MPE keyboard, so that the left split makes use of CH 1 for the common channel, while 2 through 8 are used as the Per-Note channels, and the right split uses CH 16 at the common and 9 through 15 are used for Per-Note data.

From the above scenario, you can see why it's rather exciting to have Per-Note controllers built directly into MIDI 2.0, without having to make special channel assignments or other arrangements. The direct incorporation of MPE-like capabilities gives us an extreme amount of versatility and control over Per-Note polyphony, however with a higher degree of data and performance resolution. The implication of this is not only that individual pitches can be bent individually in a polyphonic fashion, but individual polyphonic controller data can be varied within a single data stream.

MIDI 2.0 Meets VST 3

Steinberg, as a member of the MMA, were involved with MIDI 2.0's development. Given that it has been designed from the ground-up to include higher resolutions, the long-established Virtual Studio Technologies (VST) plugin protocol (in the form of VST3 instruments and other plugin types) is designed to fully make use of 2.0's expanded resolution capabilities.

MIDI CAPABILITY INQUIRY (MIDI-CI)

The secret to the above-mentioned three Bs of MIDI 2.0 lies in its ability to have a bi-directional dialog between instruments, devices and software as to what the capabilities of the involved systems are and then to configure their communications accordingly, so as to most effectively communicate. The protocol for this communication is known as MIDI Capability Inquiry (MIDI-CI) (Figure 3.3).

The added functions that MIDI 2.0 brings to devices are made possible through the use of MIDI-CI. The basic concept to this bi-directional dialog is that devices

MIDI 2.0 Environment

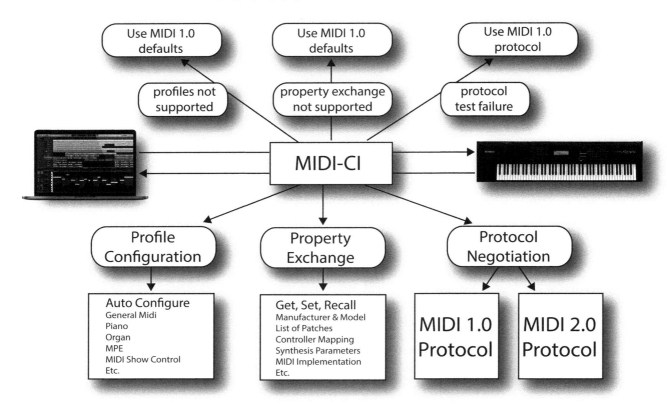

FIGURE 3.3
MIDI 2.0's MIDI-CI
environment.

can exchange information as to their basic capabilities. MIDI-CI is then used to negotiate or auto-configure any features that are common between the devices. It then provides test mechanisms that can check compatibilities when enabling these new features. If a test fails, the communication will revert to using the MIDI 1.0 protocol. In the end, the whole idea behind MIDI-CI is to expand MIDI's capabilities (such as higher resolution and Per-Note control), while protecting backwards compatibility.

THE THREE PS

MIDI-CI includes inquiries for three major areas of expanded MIDI functionality, known as "The Three Ps":

- Profile Configuration
- Property Exchange
- Protocol Negotiation

Profile Configuration

MIDI-CI Profile Configuration allows for the use of agreed conventions within the electronic music industry and manufacturers, whereby a setup profile (an agreed upon settings configuration) can be called up when needed to best configure connected hardware and software for automatic and intuitive use, thereby potentially eliminating the need for manual configuration of these devices by users.

Advanced users might be familiar with the "mapping" conventions that have been created by such companies as Native Instruments and Novation for pre-mapping the various controllers of an instrument or effects plugin over a hardware/software layout. This mapping eliminates the need to manually configure the assignments for each controller, thereby making the process seamless and automatic.

MIDI-CI is likewise capable of natively mapping all of the controllers from one device to another in an equally seamless fashion. Profiles can be written for device types or for unique applications and/or devices that could be universally adopted by the industry for a wide range of hardware/software instruments, controllers or effects control layouts. Such profiles could, of course, be used in non-musical applications, such as with robotics, industrial machines or lighting controllers.

In one example, such a MIDI controller map could be configured for a grand piano sample setup, which could be programmed with standard configurations (such as Note-On/Off, sustain pedal, etc.). However, additional parameters could also be programmed into the profile that would allow for one or more specialized velocity curves, variable sustain pedals, variable open lid angles, tuning, decay and other parameters. Any device or software system that has been designed to conform to this specific and industry-standardized profile would be automatically "mapped" throughout the system (DAW, hardware, software and controllers) to this convention.

In another example, a workstation might use MIDI-CI to query a hardware device which can be used as a hardware mixer. Once the profile is confirmed the faders, pan pots and other mixer parameters could be configured and made available to the user. Likewise, a drawbar controller could be instantly mapped to the various control parameters of an organ software plugin … or lighting control parameters could be adopted by the industry that would automatically map the needed parameters to create a software or hardware control surface in a way that can save time and eliminate tedious manual programming

Once adopted, these profiles can be used to auto-configure the settings between DAW, hardware, software and controller systems in a standardized way, allowing MIDI map profiles to pave the way for system setups that can be automatically called up without any muss or fuss.

Property Exchange

Through the use of MIDI-CI, Property Exchange allows for the access and sharing of configuration data between devices. This means that information such as parameter lists, controller auto-mapping, synth parameters and setup information about patch presets can be automatically shared. It can choose programs and patches by name and visually display relevant control and display data to DAWs without any prior device knowledge or specially crafted software.

Property Exchange makes use of JavaScript Object Notation (JSON) via System Exclusive messages for exchanging data set information and adding an extended degree of potential possibilities to MIDI 2.0. Using these messages, info such as patch and setup data can be instantly relayed between devices using their actual names and other meaningful info, which can be recognized by the user. Gone would be the use of generic numbers or abstract parameter lists that are hard to decipher in the heat of production. It could even be used to display everything you need to know about your hardware synthesizer onscreen, effectively making hardware total synth parameter recall just as possible as with software synths. This could easily turn your recall and control over "Program #32" or "Controller #9" into control over your fave patch "Dreamstate" … It's much better as a human being to know the actual name of what you're working with. Such is the power of Property Exchange.

Protocol Negotiation

MIDI-CI Protocol Negotiation allows devices to select between using the MIDI 1.0 Protocol or the MIDI 2.0 Protocol. Two devices that have established a two-way MIDI-CI session can negotiate a protocol and the various features of that protocol.

The MIDI 1.0 and the MIDI 2.0 Protocol have many messages in common, which are identical in both protocols. The MIDI 2.0 extends some of the 1.0 messages with a higher resolution and new set of features. Of course, some messages are exclusive to the MIDI 2.0 Protocol.

THE UNIVERSAL MIDI PACKET

MIDI 2.0 has a new Universal MIDI Packet format for carrying MIDI 1.0 Protocol messages and MIDI 2.0 Protocol messages. A Universal MIDI Packet (Figure 3.4) contains a MIDI message which consists of one to four 32-bit words.

The Universal MIDI Packet format is suited to sending MIDI data over high-speed transports such as USB or a network connection or between applications running inside a personal computer OS.

The traditional 5 pin DIN transport from MIDI 1.0 uses a byte stream rather than packets. At the moment, there is no plan to use the Universal MIDI Packet on the 5 pin DIN transport. Unless/until that plan changes, 5 pin DIN will only support the MIDI 1.0 Protocol.

Message Types

The first 4 bits of every message contain a Message Type (Table 3.1 and Figure 3.5). The Message Type is used as a classification of message functions.

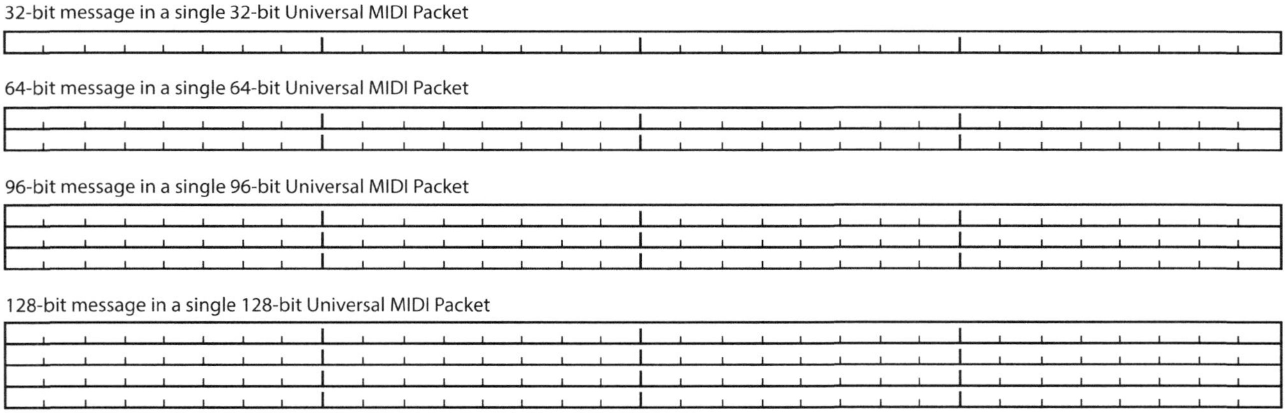

32-bit message in a single 32-bit Universal MIDI Packet

64-bit message in a single 64-bit Universal MIDI Packet

96-bit message in a single 96-bit Universal MIDI Packet

128-bit message in a single 128-bit Universal MIDI Packet

FIGURE 3.4
Universal MIDI Packet message format.

MT	Packet Size	Description
Table 3.1		Universal MIDI Packet Message Type
0x0	32 bits	Utility Messages
0x1	32 bits	System Real Time and System Common Messages (except SysEx)
0x2	32 bits	MIDI Channel Voice Messages
0x3	64 bits	Data Messages (including SysEx)
0x4	64 bits	MIDI 2.0 Channel Voice Messages
0x5	128 bits	Data Messages
0x6	32 bits	Reserved
0x7	32 bits	Reserved
0x8	64 bits	Reserved
0x9	64 bits	Reserved
0xA	64 bits	Reserved
0xB	96 bits	Reserved
0xC	96 bits	Reserved
0xD	128 bits	Reserved
0xE	128 bits	Reserved
0xF	128 bits	Reserved

Groups

The Universal MIDI Packet carries 16 groups of MIDI messages, with each group containing an independent set of system messages and 16 MIDI channels. Therefore, a single connection using the Universal MIDI Packet carries up to 16 sets of system messages over up to 256 channels.

System Real Time and System Common Messages (32-bits)

mt = 0x1	group	status	data

MIDI 2.0 Channel Voice Messages (64-bits)

mt = 0x4	group	status	index
data			

FIGURE 3.5
Universal MIDI Packet message type examples.

Each of the 16 groups can carry either MIDI 1.0 or MIDI 2.0 Protocol data. Therefore, a single connection can carry both protocols simultaneously. MIDI 1.0 Protocol and MIDI 2.0 Protocol messages, however, cannot be mixed together within one group.

Jitter Reduction Timestamps

The Universal MIDI Packet format adds a Jitter Reduction Timestamp mechanism. This Timestamp can be prepended to any MIDI 1.0 Protocol message or MIDI 2.0 Protocol message for improved timing accuracy.

MIDI 1.0 Protocol inside the Universal MIDI Packet

All existing MIDI 1.0 messages are carried in the Universal MIDI 1.0. As an example, Figure 3.6 shows how MIDI 1.0 Channel Voice Messages are carried in 32-bit packets.

FIGURE 3.6
MIDI 1.0 Channel Voice Messages within a Universal MIDI Packet.

System messages, other than System Exclusive, are encoded similarly to Channel Voice Messages. System Exclusive messages vary in size, and can be very large and can span multiple Universal MIDI Packets.

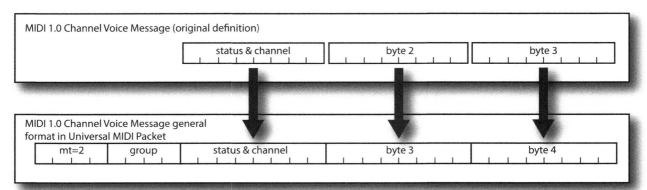

MIDI 1.0 Channel Voice Message (original definition)

status & channel	byte 2	byte 3

MIDI 1.0 Channel Voice Message general format in Universal MIDI Packet

mt=2	group	status & channel	byte 3	byte 4

Note: MIDI 1.0 Channel Voice Messages are not all 3 bytes in length. When putting shorter MIDI 1.0 messages into the Channel Voice Message General Format packet, unused bytes are reserved and set to zero.

MIDI 2.0 Protocol Messages

The MIDI 2.0 Protocol uses the architecture of the MIDI 1.0 Protocol to maintain backward compatibility and easy translation while offering expanded features.

- Extends the data resolution for all Channel Voice Messages.
- Makes some messages easier to use by aggregating combination messages into one atomic message.
- Adds new properties for several Channel Voice Messages.
- Adds several new Channel Voice Messages to provide increased Per-Note control and musical expression.
- Adds new data messages including System Exclusive 8 and Mixed Data Set. The System Exclusive 8 message is very similar to MIDI 1.0 System Exclusive but with 8-bit data format. The Mixed Data Set Message is used to transfer large data sets, including non-MIDI data.
- Keeps all System messages the same as in MIDI 1.0.

EXPANDED RESOLUTION AND EXPANDED CAPABILITIES

The example of a MIDI 2.0 Protocol Note message (Figure 3.7) shows the expansions that are used beyond their MIDI 1.0 Protocol equivalents. The MIDI 2.0 Protocol Note-On, for example, has higher velocity resolution. The two new fields, Attribute Type and Attribute data field, provide space for additional data such as articulation or tuning details.

The MIDI 2.0 Protocol replaces RPN and NRPN, having 16,384 Registered Controllers and 16,384 Assignable Controllers that are as easy to use as Control Change messages.

Creating and editing RPNs and NRPNs with the MIDI 1.0 Protocol requires the use of compound messages. These can be confusing or difficult for both developers and users. MIDI 2.0 Protocol replaces RPN and NRPN compound messages with single messages. The new Registered Controllers and Assignable Controllers are much easier to use.

Managing so many controllers might be cumbersome. Therefore, Registered Controllers (Figure 3.8) are organized in 128 Banks, with each Bank having 128 controllers. Assignable Controllers are also organized into 128 Banks, with each Bank having 128 controllers.

FIGURE 3.7
MIDI 2.0 Protocol Note message showing expanded resolution and capabilities.

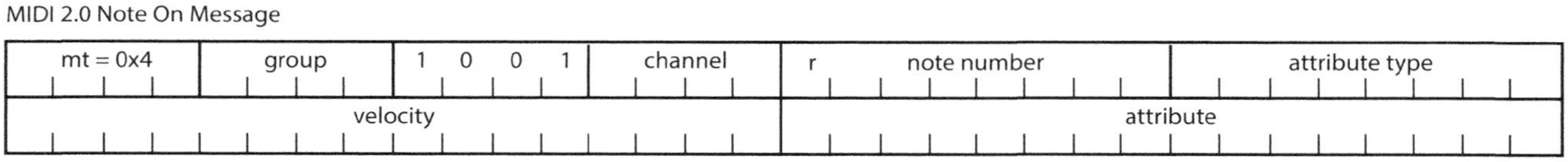

MIDI 2.0 Note On Message

mt = 0x4	group	1 0 0 1	channel	r	note number	attribute type
velocity			attribute			

MIDI 2.0 Registered Controller Message

mt = 0x4	group	0 0 1 0	channel	r	bank	r	index
data							

MIDI 2.0 Assignable Controller Message

mt = 0x4	group	0 0 1 1	channel	r	bank	r	index
data							

FIGURE 3.8
Registered Controllers and Assignable Controllers support data values up to 32 bits in resolution.

MIDI 2.0 Program Change Message

The MIDI 2.0 Protocol combines the Program Change and Bank Select mechanisms from the MIDI 1.0 Protocol into one message (Figure 3.9). The MIDI 1.0 mechanism for selecting Banks and Programs requires sending three MIDI messages. MIDI 2.0 changes this mechanism by combining the Bank Select and Program Change into one new MIDI 2.0 Program Change message.

The MIDI 2.0 Program Change message always selects a Program. A Bank Valid bit (B) determines whether a Bank Select is also performed by the message. If Bank Valid = 0, then the receiver performs the Program Change without selecting a new Bank and the receiver keeps its currently selected Bank. Bank MSB and Bank LSB data fields are filled with zeroes. If Bank Valid = 1, then the receiver performs both Bank and Program Change. Other option flags that are not yet defined and are Reserved for future use.

New Data Messages for MIDI 1.0 and MIDI 2.0 Protocol

New data messages include both a System Exclusive 8 and Mixed Data Set. The System Exclusive 8 message is very similar to MIDI 1.0 System Exclusive but with an 8-bit data format. The Mixed Data Set message is used to transfer large data sets, including non-MIDI data. Both messages can be used when using the Universal MIDI Packet format for the MIDI 1.0 or MIDI 2.0 Protocols.

The Future of MIDI 1.0

MIDI 1.0 is not being replaced. Rather it's being extended and is expected to continue its integration into the new MIDI 2.0 environment. It is part of the Universal MIDI Packet, the fundamental MIDI data format. Many MIDI devices will not need any or all of the new features of MIDI 2.0 in order to perform all

FIGURE 3.9
MIDI 2.0 Program Change Message.

MIDI 2.0 Registered Controller Message

mt = 0x4	group	1 1 0 0	channel	reserved	open options
r program		reserved	r bank msb	r bank lsb	

their functions. Some devices will continue to use the MIDI 1.0 Protocol while using other extensions of MIDI 2.0, such as Profile Configuration or Property Exchange.

The Future of MIDI 2.0

Obviously, MIDI 2.0 is very much in its infancy, as of this writing. For further information and updated information as the spec, instrument and usage of 2.0 begin to develop, it's best to follow the changes and expanded capabilities of the spec that are posted on the MIDI Manufacturer's Association website (www. midi.org).

CHAPTER 4

Systems Interconnections

From Chapter 1, we know that MIDI is a digital language that's used to communicate control information between connected devices within music or other production systems. As such, there are numerous ways that this data can be communicated within a connected system; these include:

- Direct wired connections using dedicated MIDI cables
- Wired communication over USB and other cabled connections
- Communication over the Web or other ISP systems
- Direct wireless communication

THE MIDI CABLE

The original method for communicating data between connected devices in a system, which of course is still in common use today, uses MIDI cables to get the job done.

A single MIDI cable is used to transmit 16 discrete channels of performance, controller and timing information in one direction, using data densities that are economically small and easily to manage. In this way, it's possible for cables to be "chained" together in a way that can easily communicate MIDI messages from a specific source (such as a keyboard or MIDI sequencer) to any number of devices within a simple, connected MIDI network (Figure 4.1). In addition, MIDI is flexible enough that multiple MIDI cable networks can used to interconnect devices in a wide range of configurations over 32, 48, 128 or more discrete MIDI channels.

A MIDI cable (Figure 4.2) consists of a shielded, twisted pair of conductor wires that has a male five-pin DIN plug located at each of its ends. The MIDI specification currently uses only three of the five pins, with pins 4 and 5 being used as conductors for MIDI data and pin 2 to connect the cable's shield to equipment ground. Pins 1 and 3 are currently not in use, although the next section describes an ingenious (but rarely used) system for powering devices through these pins

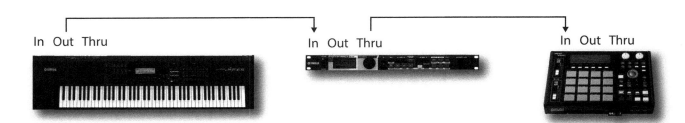

In   Out   Thru   In   Out   Thru   In   Out   Thru

FIGURE 4.1
Simple MIDI cable network.

that's known as MIDI phantom power. The cable system uses twisted wire and metal shield groundings to reduce outside interference, such as radio frequency interference (RFI) or electrostatic interference, both of which can serve to distort or disrupt the transmission of MIDI messages.

MIDI pin description:

- Pin 1 is not used in most cases; however, it can be used to provide the V – (ground return) of a MIDI phantom power supply.
- Pin 2 is connected to the shield or ground cable, which protects the signal from radio and electromagnetic interference.
- Pin 3 is not used in most cases; however, it can be used to provide the V+ (+ 9 to +15 V) of a MIDI phantom power supply.
- Pin 4 is a MIDI data line.
- Pin 5 is a MIDI data line.

MIDI cables come pre-fabricated in lengths of 2, 6, 10, 20 and 50 feet. To reduce signal degradations and external interference that tend to occur over extended cable runs, 50 feet is the maximum length that's specified by the MIDI spec.

FIGURE 4.2
The MIDI cable: (a) wiring diagram; (b) five-pin DIN connectors.

MIDI JACKS

MIDI is distributed from device to device using three types of MIDI jacks: MIDI In, MIDI Out and MIDI Thru (Figure 4.3a). These three connectors use five-pin

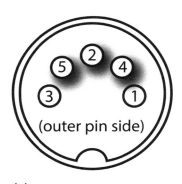

pin 1 - no connection
pin 4 - MIDI signal
pin 2 - ground
pin 5 - MIDI signal
pin 3 - no connection

(a)

(b)

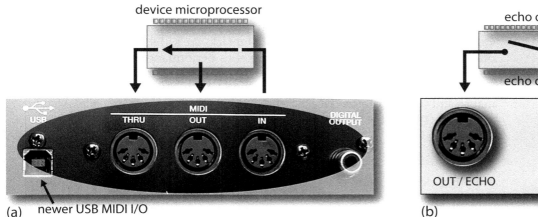

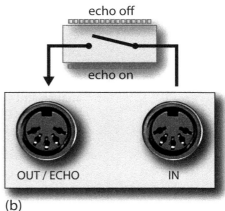

(a) newer USB MIDI I/O (b)

DIN jacks as a way to connect MIDI instruments, devices and computers together within a production system. As a side note, it's nice to know that these ports (as strictly defined by the MIDI 1.0 spec) are optically isolated to eliminate possible ground loops that might occur when connecting numerous devices together.

FIGURE 4.3
MIDI hardware ports. (a) MIDI In, Out and Thru ports, showing the device's various signal path routing; (b) MIDI echo on/off routing.

MIDI In Jack

The MIDI In jack receives messages from an external source and communicates this performance, control and timing data to the device's internal microprocessor, allowing an instrument to be played or a device to be controlled from a source.

MIDI Out Jack

The MIDI Out jack is used to transmit MIDI performance or control messages from one device out to another MIDI instrument or device.

MIDI Thru Jack

The MIDI Thru jack retransmits an exact copy of the data that's being received at the MIDI In jack. This process is important, because it allows data to pass directly through an instrument or device to the next device within the MIDI chain. Keep in mind that this jack is used to relay "only" an exact copy of the data that's coming in from the MIDI In port and generally isn't merged with data being transmitted from the MIDI Out jack.

MIDI Echo

Certain early MIDI devices may not include a MIDI Thru jack at all. These devices, however, will usually give the option of switching the MIDI Out between being an actual MIDI Out jack and a MIDI Echo jack (Figure 4.3b). As with the MIDI

Thru jack, a MIDI Echo option can be used to retransmit an exact copy of any information that's received at the MIDI In port and route this data to the MIDI Out/Echo jack. Unlike a dedicated MIDI Out jack, the MIDI Echo function can often be selected to merge incoming data with performance data that's being generated by the device itself. In this way, more than one controller can be placed in a MIDI system at one time. It should be noted that, although performance and timing data can be echoed to a MIDI Out/Echo jack, not all devices are capable of echoing SysEx data.

THE DAISY CHAIN

Although electronic studio production equipment and setups are rarely alike (or even similar), there are a number of general rules that make it easy for MIDI devices to be connected into a functional network. These common configurations allow MIDI data to be distributed in the most efficient and understandable manner possible.

As a general rule, there are only two valid ways to connect one MIDI device to another within a MIDI chain (Figure 4.4):

- Connecting the MIDI Out jack of a source device (controller or sequencer/computer) to the MIDI In of a second device in the chain.
- Connecting the MIDI Thru jack of the second device to the MIDI In jack of the third device in the chain, and following this same Thru-to-In convention until the end of the chain is reached.

One of the simplest and most common ways to distribute data throughout a cabled MIDI system is the *daisy chain*. This method relays MIDI data from a first (source) device (controller or sequencer/computer) to the MIDI In jack of the second device in the chain (which receives and acts upon this data). This second device then relays an exact copy of the incoming data out to its MIDI Thru jack, which is then relayed to the third device in the chain, and so on through the successive devices. In this way, up to 16 channels of MIDI data can be chained from one device to the next within a connected data network – and it's precisely this concept of transmitting multiple channels through a single MIDI line that

FIGURE 4.4
The two valid means of connecting one MIDI device to another via standard hardware cables.

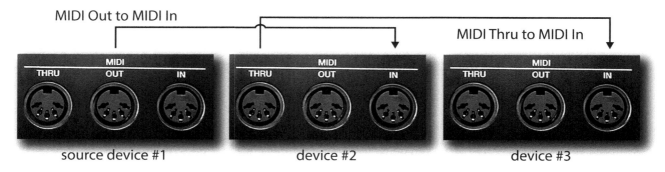

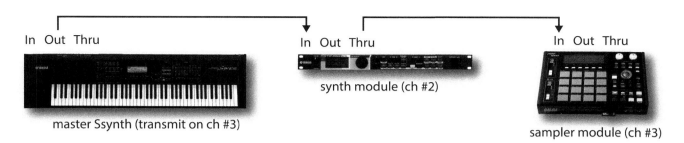

In Out Thru

master Ssynth (transmit on ch #3)

In Out Thru

synth module (ch #2)

In Out Thru

sampler module (ch #3)

makes this simple concept work! Let's try to understand this system better by looking at a few examples.

Figure 4.5 shows a simple (and common) example of a MIDI daisy chain whereby data flow from a controller (MIDI Out jack of the source device) to a synth module (MIDI In jack of the second device in the chain). An exact copy of the data that flow into the second device is then relayed to its MIDI Thru jack to a sampler (MIDI In jack of the third device in the chain). From the section on MIDI channels in Chapter 2, it shouldn't be hard to understand that if our controller is set to transmit on MIDI channel 3 the second synth in the chain (which is set to channel 2) will ignore the messages and not play, while the third device (which is set to channel 3) will be playing its heart out. The moral of this story is that, although there's only one connected data line, a wide range of instruments and channel voices can be played in a surprising number of combinations – all by assigning channel numbers to instruments and devices along a daisy chain.

FIGURE 4.5
Example of how a typical daisy chain setup can be used to transmit performance data to various instruments in the chain by assigning channel numbers.

The Multiport Network

Another common approach to routing wired MIDI throughout a production system involves distributing MIDI data through the multiple 2-, 4- and 8-In/Out ports that are available on the newer multiport MIDI interfaces or through the use of multiple USB MIDI interfaces.

In larger, more complex MIDI systems, a multiport MIDI network offers several advantages over a single daisy chain path. One of the most important is its ability to address devices within a complex setup that requires more than 16 MIDI channels. For example, a 2 × 2 MIDI interface that has two independent In/Out paths is capable of simultaneously addressing up to 32 channels (i.e., Port A 1–16 and Port B 1–16), whereas an 8 × 8 port is capable of addressing up to 128 individual MIDI channels.

A multiport setup can also be used to designate a MIDI controller as the master source in the system, so a DAW sequencer can be used to control the entire playback and channel routing functions within a separate daisy-chain (Figure 4.6). In this situation, the MIDI data flow from a master controller to the MIDI In jack of a computer's MIDI interface. This data is then routed directly to the MIDI functions on the DAW, where it is then routed out from the B port to the first instrument in the MIDI chain (which is, for example, set to receive on Port B,

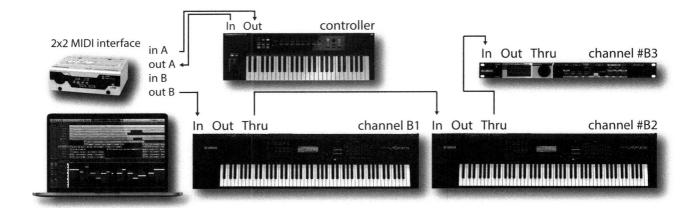

FIGURE 4.6
Example of how a MIDI keyboard controller can be connected to a DAW system in a multiport daisy chain fashion, that can be used to transmit performance data to various instruments in the chain by easily and automatically routing channel numbers from with the DAW/ sequencer session.

MIDI channel 1). The data chain is then routed thru to the MIDI In of the next instrument in the chain (which is set to Port B, Channel 2), and thru to the next in the chain (Port B, channel 3) … and so on.

When we stop and think about it, we can see that the controller is essentially used as a central "performance tool" for integrating with the DAW for entering performance and controller data into the MIDI sequencer, which is then used to communicate out to the various instruments throughout the connected MIDI chain in a relatively straightforward fashion.

Note: although the distinction isn't overly important, you might want to keep in mind that a MIDI "port" is a virtual data path that's processed through a computer, whereas a MIDI "jack" is the physical connection to the device itself.

This type of network of independent MIDI chains has a number of advantages. As an example, Port A might be dedicated to three instruments that are set to respond to MIDI channels 1 to 6, channel 7, and finally channel 11, whereas Port B might be transmitting data to two instruments that are responding to channels 1–4 and 5–10 and Port C might be communicating SysEx MIDI data to and from a MIDI remote controller for a digital audio workstation (DAW).

In this modern age of audio interfaces, multiport MIDI interfaces and controller devices that are each fitted with USB and physical MIDI ports can be readily available to a connected system. Using USB and other data protocols, it's a simple matter for a computer to route and synchronously communicate MIDI data throughout the studio in any number of ingenious and cost-effective ways, without the need for dedicated MIDI cables.

ALTERNATIVE WAYS TO CONNECT MIDI TO THE PERIPHERAL WORLD

An important event in the evolution of personal computing has been the maturation of hardware and processing peripherals. With the development of FireWire™ (www.1394ta.org), USB (www.usb.org) and Thunderbolt (https://thunderbolt technology.net), hardware devices such as mice, keyboards, cameras, audio interfaces, MIDI interfaces, instruments, drives and even portable fans can be plugged into an available port without any frustrating need to change hardware settings.

The use of these communications protocols has changed the face of the MIDI studio in a number of ways, the greatest being the ability to directly communicate MIDI data over protocol lines from a host to a connected device in a bi-directional fashion, without the need for dedicated MIDI cables.

By far, the most common way to connect peripheral MIDI devices together is through USB, whose initial spec goals were:

- Standardize connector types: there are obviously a wide range of connector types, however, they are all fairly standardized, adaptable and use the same USB data protocol.
- Hot-swappable: USB devices can be safely plugged and unplugged as needed while the computer is running. So there is no need to reboot.
- Plug and play: USB devices are divided into functional types (audio, image, human user interface, mass storage). The operating system software will often automatically identify, configure and load the appropriate device driver when a user connects a USB device.
- Fast connection speeds: USB offers low-speed (1.5 Mbit/s), full-speed (12 Mbit/s) and high-speed (up to 480 Mbit/s) transfer rates that can support a variety of USB peripherals. USB 3.0 (SuperSpeed USB) achieves the throughput up to 5.0 Gbit/s, while USB 3.1 or USB-C can even work at speeds up to 10 G/sec.
- Expandability: Up to 127 different peripheral devices may theoretically be connected to a single bus at one time

USB System Architecture

The basic USB system architecture is actually pretty simple and consists of the following main components (Figure 4.7):

- A "host" computer, smartphone or tablet
- One or more connected USB "devices"
- A physical bus represented by the USB cable that links the devices with the host

It's important to know that all data transfers from the host to a device or between devices are initiated and controlled by the host and that USB peripherals are slaves that respond to the host's instructions. Therefore, when connecting USB

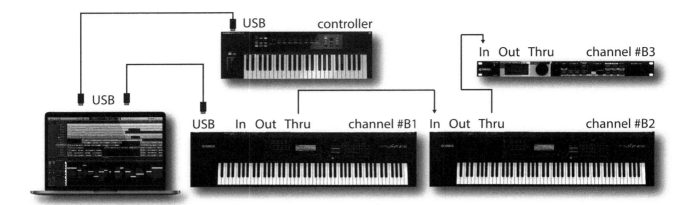

USB   controller

In Out Thru   channel #B3

USB

USB   In Out Thru   channel #B1   In Out Thru   channel #B2

FIGURE 4.7
The basic USB system architecture is connected from a
"host" (in this case, a laptop)
to one or more connected
"devices".

MIDI peripheral devices you'll need a computer, smartphone or tablet in the system in order to control and initiate USB communication. This allows data to be communicated from the device to the host, from the host to a connected device or from one device, through the host and to another device in the system.

USB has opened up a wide range of interconnection possibilities, for example:

- A simple and inexpensive USB MIDI interface can be used to integrate a single MIDI chain into your production system.
- An audio interface might include MIDI I/O ports that can be connected to a MIDI chain.
- A multiport USB MIDI interface can be used to offer up multiple instrument/device chains into your system.
- A MIDI instrument or controller might offer one or more MIDI I/O ports that can be used to connect instruments or devices into your system.

For those getting started, these useful and cost-saving options make it relatively easy to integrate your existing instruments and devices into your DAW and sequencing environment.

Wireless MIDI

A small number of companies have begun to manufacture wireless MIDI transmitters (Figure 4.8) that allow keyboards, MIDI guitars, wind controllers, etc., to be footloose and fancy free on-stage and in the studio. Working at maximum distances that can vary between manufacturers, these battery-powered transmitter/receiver systems introduce low-delay latencies and can be transmitted from a device to its host via Bluetooth or over user-selectable radio frequencies.

THE MIDI INTERFACE IN REAL-WORLD APPLICATIONS

Although computers and electronic instruments both communicate using the digital language of 1s and 0s, computers simply can't understand the language

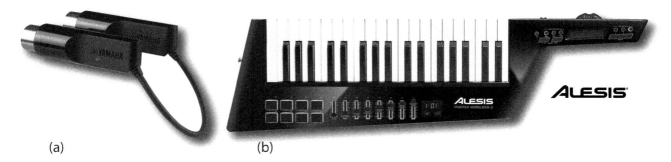

(a)   (b)

FIGURE 4.8
Wireless MIDI transmitter systems. (a) Yamaha Wireless MD-BT01 5-PIN DIN MIDI Adapter (courtesy of Yamaha Corporation of America; www.yamaha.com); (b) VORTEX WIRELESS 2 Wireless USB/MIDI Keytar Controller. (Courtesy of alesis; www.alesis.com.)

of MIDI without the use of a device that translates the serial messages into a data structure that computers can comprehend. Such a device is known as a *MIDI interface*. A wide range of MIDI interfaces currently exists that can be used with most computer systems and OS platforms. For the casual and professional musician, interfacing MIDI into a production system can be done in a number of ways.

- Through a dedicated MIDI interface that can then be wired directly to instruments or devices in a connected network via standard MIDI cables.
- Through a device that can communicate throughout a connected network, through the host CPU via FireWire™, USB or Thunderbolt connection or controller surface.

Understanding how the physical and virtual connections of MIDI are made in a system can be a bit daunting for the budding artist; therefore, now seems as good a time as any to walk through several possible routing scenarios and see how things can be logically thought "thru" to get your musical rig up and running.

Another Word about Audio Inputs

It is worth mentioning again that from Chapter 1 that MIDI and audio will almost always operate using separate paths, whether they are physical or virtual in nature. Of course, virtual instruments will always be able to route both their MIDI and audio signal paths internally in the digital domain. If a studio is full of older hardware gear, the MIDI and audio connections must be physically made. Audio paths within a studio can be connected in any of several ways:

1. The audio can be individually patched or plugged into your audio interface, where the audio (often a stereo pair) can be recorded and imported into a session's audio track. This straightforward system can help reduce system complexity and clutter through the simple use of a cable or a patchbay system that lets you patch between instruments.
2. You could set up a situation whereby you plug your various instruments into a hardware or summing mixer that could route all of your instruments to a single interface.
3. Virtual plugin instruments could be digitally routed to a mixer window track(s) within the DAW.

4. Lastly, you could get an audio interface that has a large number of audio inputs that let you connect your music gear directly into the DAW for mixing, routing and recording.

Each of these approaches has its merits and the choice will depend upon your preferred working style.

Simple Daisy Chain Routing

Let's start our understanding of MIDI daisy chain routing by looking at the most common of setups, one that makes use of a laptop and several hardware devices which are connected to the host computer via standard MIDI cables. As we know from Chapter 2, whenever a MIDI device or sound generator within a device or program function is instructed to respond to a specific channel number, it will only react to messages that are transmitted on that channel (i.e., it ignores all channel messages that are transmitted on any other channel). Now, let's go about the task of creating a short song using the laptop/DAW that is connected via a USB MIDI interface to a keyboard synth and two other instruments, as shown in Figure 4.9. Here, I'm going to assume that you know your basic way around your DAW/sequencer and its settings.

1. If we created a setup that's much like Figure 4.9, we could open up a new session on the DAW and then creating a MIDI track that's set to transmit on channel 10 and receive data from all MIDI channels. Since the first keyboard controller/synth in the chain is capable of playing drum/percussion sounds, let's set it to a favorite drum set patch and set that instrument to respond to channel 10. By activating the input monitor button on our track, we should be able to play the keyboard and listen to what's being played (assuming that you've routed all of the instrument audio outs to a mixer or inputs on your audio interface). Now, we can finally start making some music.

2. We could then press the music keys and begin to play the kick and the snare parts. Once ready, we can put the track into record and lay down a 4-bar kick/snare beat. Since the chain allows all MIDI data/channel information to pass from one instrument thru to the next, the data will be passed through to the entire chain, back to the MIDI interface In port – where it is recorded to the track. Once the beat has been made, we could loop the beat and listen to our work.

3. Next, we'll create a new MIDI track and set it to transmit on channel 2. Since the second synth in the chain is set to respond to channel 2, we can activate the track's input monitor and audition its sounds. Playing the sequence will begin playing our drum beats on the first synth and we can practice our riff on the second synth. Once ready, we can lay down that track. Since, the data on the first track is transmitting on channel 10 to the first synth and the second track is now sending on channel 2 to the second synth, and we can now hear the song begin to take shape.

4. Next, we can create a third track that is set to transmit on MIDI channel 3 to the next instrument and we can go about practicing and then laying down a third instrument track.

5. If you want to, you could then create a fourth track that could be set to transmit on another MIDI channel that could then be received by the first synth (if it's capable of playing multiple sounds at once) … allowing you to keep on building on your idea. Of course the idea here is to improvise, learn and have fun.

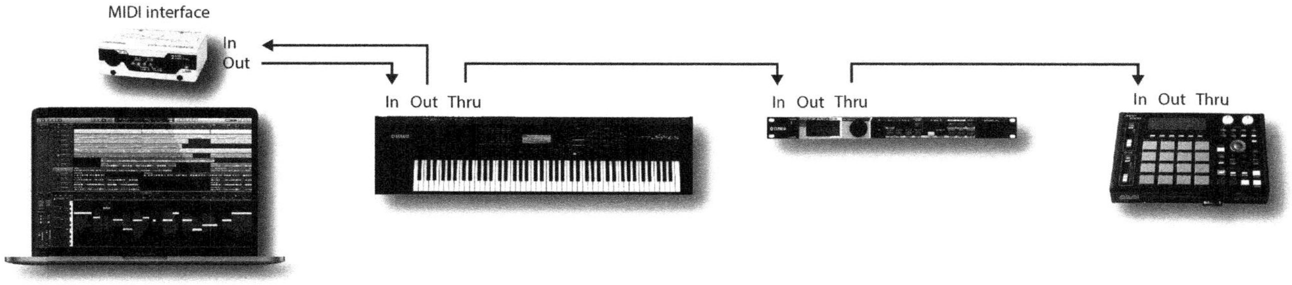

It goes without saying that the above simple example is just one of an infinite setup and channel possibilities that can be encountered in a production environment. It's often true, however, that even the most complex MIDI and production rooms will have a system strategy – a basic channel and overall layout that makes the day-to-day operation of making music easier. This layout and the basic decisions that you might make in your own room are, of course, up to you. Streamlining a system to work both efficiently and easily will come with time, experience and practice.

FIGURE 4.9
Example of a basic MIDI setup.

Multiple and Multiport MIDI Interface Routing

A system that makes use of multiple MIDI interface routing chains or a single interface which has multiple I/O ports could be set up much the same as the single chain example that is shown above. Obviously, the difference is that several chains can be built up in a system that's connected to a single host device.

There can be several reasons for having a multiple port MIDI chain within a system, for example:

- The chosen controller could be assigned to its own data port path. This would keep the main studio controller in a simple and isolated MIDI In and Out setup which is then routed directly to and from the DAW.
- Instruments within a studio might be physically separated from each other. One group of instruments might be located on the left-hand side of the studio, and this group might be chained to MIDI interface A. Another set of instruments to the right could then be chained into MIDI interface B. This keeps the setup simple and straightforward in the studio without the need for running dedicated MIDI cables all around the room. It can also reduce possible latencies that might be introduced by having too many instruments on a single chain.
- If an older MIDI and/or DAW controller (one that only offers direct MIDI cable I/O and not USB connectivity) is used in a single chain, the data throughput might be too high for the chain and might introduce delays or other artifacts that could clog the line. This is the perfect excuse to place any controllers on a separate MIDI chain. Of course, newer devices that use USB will generally connect directly through the USB data line to the host device and any problems with data clogging are usually eliminated.

Obviously, the world of MIDI has changed as we have all adapted to using computers and peripheral devices and instruments that connect directly to the host computer via the serial communication protocols of Firewire, USB and Thunderbolt. These forms of "virtual routing" obviously differ from actual cable routing, in that the MIDI data is carried alongside audio and other data formats; however, it's important to keep in mind that these multiple lines still strictly adhere to the MIDI 1.0 and 2.0 protocols. The concept that the data flow from a single source to a destination and then out to another virtual destination should always be kept in mind when instruments, processors or plugins are routed using dialog boxes instead of cables. The routing connections may be virtual, but they follow the same connection logic as they would with physical cable connections.

In order to understand virtual connections, let's walk through another DIY tutorial, so as to best learn the process and make the connections for ourselves.

1. Let's start by opening up a new or existing session and then adding a new MIDI track. Here, we're going to connect our controller's In and Out ports to the Out and In ports on a multiport interface.

2. In the I/O dialog box on the track, we're going to set our MIDI controller port and channel as the source. This could even be made easier by telling the track to respond to "all" incoming MIDI channels. Since the controller is the only device that will transmit to the DAW, this will automatically, always be your selected input.

3. We can now route the second MIDI I/O port to one or more instruments by selecting the MIDI Out port and channel routing dialogs to correspond to the correct instruments and channels.

4. By simply assigning the tracks to their corresponding ports and tracks, this part of the process should be relatively straightforward.

5. The next part of the puzzle would be to connect the instrument's audio outputs to your audio interface. This could be done by simply plugging the instrument into two input channels on your two-channel interface, or by creating a more complex audio routing scheme via a separate audio mixer of multi-input audio interface.

6. By making sure that your tracks are set to record ready and that the inputs are set to monitor the sources, you should now be able to play the keyboard and hear the instrument's output through the system.

Virtual Plugin Routing

In this day and age, there are as many virtual plugin instruments (Figure 4.10) as there are hardware ones … probably far more. The nice thing about inserting a plugin into a session is that often the audio connections (and sometimes even that MIDI routings) are automatically made by the DAW itself.

Next, let's try adding a software plugin instrument to our session using a single instrument track.

1. If your DAW will allow it, let's add an "instrument" track to the session. In this way, we can go about easily inserting a software instrument plugin into the project.

2. Next, we can select the desired software instrument and then go about choosing a MIDI source (such as a MIDI controller). If the instrument and the MIDI tracks must be made separately, create both the instrument and the MIDI tracks and then go about the task of assigning the MIDI Out to that plugin.

3. By placing the track into record and enabling the input source (on both the instrument and MIDI tracks, if that's needed by the DAW software), we should be able to play the keyboard and hear the instrument as it's being played.

Of course, not all DAWs are deigned to work the same and you may have to consult your manual in order to get everything working, but with patience and determination, you should be up and running without too much difficulty.

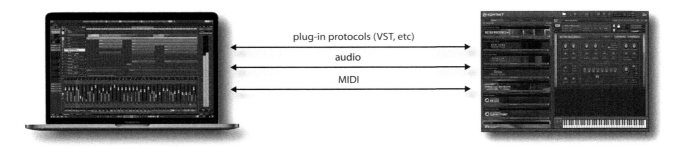

plug-in protocols (VST, etc)

audio

MIDI

FIGURE 4.10
Example of a virtual instrument within a session.

CHAPTER 5

Electronic Instruments

Since their inception in the early 1980s, MIDI-based electronic instruments have played an important role in the development of music technology and production. These devices, along with the advent of cost-effective analog and digital audio recording systems, have been amongst the most important technological advances to shape the industry into what it is today. In fact, the combination of these technologies has turned the personal project studio into one of the most important driving forces behind modern-day music production.

INSIDE THE TOYS

Although electronic instruments often differ from one another in looks, form and function, they almost always share a common set of basic building block components (Figure 5.1), including the following:

- Central processing unit (CPU) – CPUs are one or more dedicated computing devices (often in the form of a specially manufactured microprocessor chip) that contain all of the necessary instructional brains to control the hardware, voice data and sound-generating capabilities of the entire instrument or device.
- Performance controllers – these include such interface devices as music keyboards, knobs, buttons, drum pads and/or wind controllers for inputting performance data directly into the electronic instrument in real time or for transforming a performance into MIDI messages. Not all instruments have a built-in controller. These devices (commonly known as modules) contain all the necessary processing and sound-generating circuitry; however, the idea is to save space in a cramped studio by eliminating redundant keyboards or other controller surfaces.
- Control panel – the control panel is the all-important human interface of data entry controls and display panels that let you select and edit sounds and route and mix output signals, as well as control the instrument's basic operating functions.

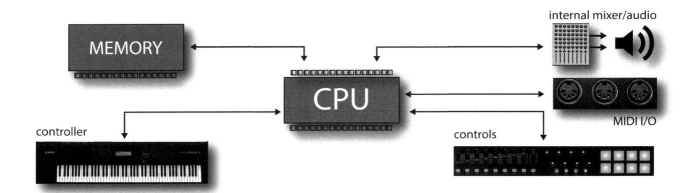

controller

controls

MIDI I/O

FIGURE 5.1
The basic building blocks of an electronic musical instrument.

- Memory – digital memory is used for storing important internal data (such as patch information, setup configurations and/or digital waveform data). These digital data can be encoded in the form of either read-only memory (ROM; data that can only be retrieved from a factory-encoded chip, cartridge or CD/DVD-ROM) or random-access memory (RAM; memory that can be read from and stored to a device's resident memory, cartridge, hard disk or recordable media).

- Voice circuitry – depending on the device type, this section can chain together digital processing "blocks" to either generate sounds (voices) or process and reproduce digital samples that are recorded into memory for playback according to a specific set of parameters. In short, it's used to generate or reproduce a sound patch, which can then be processed, amplified and heard via speakers or headphones.

- Auxiliary controllers – these external controlling devices (not shown in figure) can be used in conjunction with an instrument or controller. Examples of these include foot pedals (providing continuous-controller data), breath controllers and pitch-bend or modulation wheels. Some of these controllers are continuous in nature, while others exist as a switching function that can be turned on and off. Examples of the latter include sustain pedals and vibrato switches.

- Communications ports – these data ports and physical MIDI jacks are used to transmit and/or receive MIDI data.

Generally, no direct link is made between each of these functional blocks; the data from each of these components is routed and processed through the instrument's CPU. For example, should you wish to select a certain sound patch from the instrument's control panel, the control panel instructs the CPU to recall all of the waveform and sound-patch parameters from memory that are associated with that particular sound. These instructional parameters modify the internal voice circuitry, so that when a key on the keyboard is pressed, the sound generators will output the desired patch's note and level values.

Of course, most if not all of an instrument's functional components can be used to access or communicate performance, patch and system setup information via MIDI. A few of the many examples include:

- Transmission of performance and control-related data between devices, throughout a connected network.
- Transmission of real-time control parameters via MIDI and/or system exclusive (SysEx) messages.
- Bulk transmission of device patch and system parameters via system exclusive (SysEx) messages.

The remainder of this chapter will focus on the various types of MIDI instruments that are currently available on the market. These devices can be grouped into such categories as keyboards, percussion, MIDI guitars and strings, woodwind instruments and almost anything else you can think of.

It almost goes without saying that, over the years, model numbers, manufacturers, communications protocols, system applications and production styles will come and go. The job of any self-respecting music professional and technogeek will be to keep up with the latest developments as technology progresses, changes and goes in and out of style. Keeping your mind open to new ways of working, through reading, surfing the Web and getting your hands experimentally dirty is a sure-fire way to stay fresh and youthful in the passionate pursuit of your career or hobby.

By far, the most common instruments that you'll encounter in almost any MIDI production facility will probably belong to the keyboard family. This is due, in part, to the fact that keyboards were the first electronic music devices to gain wide acceptance; also, MIDI was initially developed to record and control many of their performance and control parameters. The two basic keyboard-based instruments are the synthesizer and the digital sampler.

THE SYNTH

A synthesizer (or synth) is an electronic instrument that uses multiple sound generators, filters and oscillator blocks to create complex waveforms that can be combined into countless sonic variations. These synthesized sounds have become a basic staple of modern music and range from those that sound "cheesy" to ones that realistically mimic traditional instruments … all the way to those that generate other-worldly, ethereal sounds that literally defy classification.

Synthesizers use a number of different technologies or program algorithms to generate sounds, such as:

- FM synthesis: this technique generally makes use of at least two signal generators (commonly referred to as "operators") to create and modify a voice (patch sound). It often does this by generating a signal that modulates or changes the tonal and amplitude characteristics of a base carrier signal.

More sophisticated FM synths use up to four or six operators per voice, each using filters and variable amplifier types to alter a signal's characteristics into a sonic voice that can be complex in nature.

- Wavetable synthesis: this technique works by storing small segments of digitally sampled sound into a memory media. Various sample-based and synthesis techniques then make use of looping, mathematical interpolation, pitch shifting and digital filtering to create extended and richly textured sounds that use a surprisingly small amount of sample memory, allowing hundreds if not thousands of samples and sound variations to be stored in a single device or program. These sample-based systems are often called wavetable synthesizers because the prerecorded samples that are encoded within the instrument's memory can be thought of as a "table" of sound waveforms that can be looked up and used when needed. Once selected, a vast range of modification parameters (such as sample mixing, envelope, pitch, volume, panning and modulation) can be called up from the device's patch memory to control a sample's overall sonic character.

- Additive synthesis: this technique makes use of combined waveforms that are generated, mixed and varied in level over time to create new timbres that are composed of multiple and complex harmonics. Subtractive synthesis makes extensive use of filtering to alter and subtract overtones from a generated waveform (or series of waveforms).

- Subtractive synthesis: this method (which can be used in combination with additive synthesis) makes extensive use of filtering to alter and subtract overtones from a generated waveform (or series of waveforms). For example, such a device could start with a square or sawtooth waveform that, with the use of filters, could be altered to approximate an acoustic instrument. These generated sounds can also be filtered and changed in level over time to more closely approximate a desired sound.

FIGURE 5.2
Roland Fantom-8 Synth. (Courtesy of Roland Corporation U.S., www.roland.com.)

Of course, synths come in all shapes and sizes and use a wide range of patented synthesis techniques for generating and shaping complex waveforms in a polyphonic fashion using 16, 32 or even 64 simultaneous voices (Figures 5.2 and 5.3).

In addition, many synths often include a percussion section that can play a full range of drum and "perc" sounds in a number of styles. Reverb and other basic effects are also commonly built into the architecture of these devices, reducing the need for using extensive outboard effects when being played on-stage or out of the box. Speaking of "out of the box", a number of synth systems are referred to as being "workstations". Such beasties are designed (at least in theory) to handle many of your basic production needs (including basic sound generation, MIDI sequencing, effects, etc.) ... all in one package.

FIGURE 5.3
Getting into the music with a Yamaha keyboard Workstation. (Courtesy of Yamaha Corporation of America, www.yamaha.com.)

Rack Synth

Synthesizers are also commonly designed into full 19-inch or half-rack mountable systems (Figures 5.4 and 5.5). These devices, which are known as rack synths or synth modules, often contain all of the features of a standard synthesizer, except they don't have a keyboard controller. This space-saving feature means that more synths can be placed into your system rack and can be controlled from a master keyboard controller and sequencer without cluttering up your system with redundant keyboards.

Software Synth

Since wavetable and other types of synths derive their sounds from prerecorded samples and/or general program language, it logically follows that these synths

FIGURE 5.4
Roland Integra-7 Synth. (Courtesy of Roland Corporation U.S., www.roland.com.)

can be easily created to operate as a fully functional software musical instrument. Software synths (Figure 5.6) can mimic literally all types of synth designs.
They can also be modeled to match the design of an original hardware unit,
often with capabilities that far outreach those of the original device.

Although most software synths are fully formed programs, plugins or apps,
some system types (Figure 5.7) allows a sound-generating or wavetable synth
to be designed using on-screen, digital building blocks. This modular, building-
block approach works by linking various signal-processing modules in a chain
or parallel fashion to generate or modify a sound. These modules consist of
such traditional synthesis blocks as oscillators, voltage-controlled amplifiers
and voltage-controlled filters. These can then be mixed and processed to alter
the signal's overall content and harmonic structure into almost any texture or
synthesized sound that could possibly be imagined. Because the system exists in
software, a newly created sound patch can obviously be saved to disk for later
recall.

As you might expect, the depth and capabilities of any software synth depends
upon the quality of the program and its generation techniques, wavetable signal
quality, sample rate and overall processing techniques.

A Synth in Every Pocket

Although it would be easy to overlook these small software wonders, by far the
greatest number of installed synthesizers are sitting on your desk or in your
pocket right now. I'm referring to the software synths that are built into every PC,
laptop and mobile phone. These synths enable the devices to play MIDI music,
ringtones, game themes and the like using a soundset format called SoundFont.

(a)

(b)

 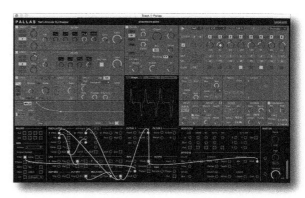

Originally developed in the early 1990s by E-mu Systems and Creative Labs, this file format makes use of wavetable and perceptually additive synthesis to layer sounds together with true stereo sample support while making efficient use of system memory. Since SoundFont is used in commonly found media systems, it naturally conforms to the General MIDI specification, which conforms to a standard patch and drum-sound structure. This allows all MIDI files to be uniformly played by any system with the correct instrument voicing and levels. Further information about General MIDI can be found within Chapter 12.

FIGURE 5.7
Ableton Max for Live (showing Pallas' MaxforCats) is a building block platform for creating your own instruments and effects, tools for live performance and visuals within Ableton Live. (Courtesy of Ableton AG, www.ableton.com.)

THE SAMPLER

A sampler is a device that can convert audio into a digital form that is then imported into internal random-access memory (RAM). Once audio has been sampled and loaded into RAM (from disk or solid-state memory), segments of sampled audio can then be edited, transposed, processed and played in a polyphonic, musical fashion. In short, a sampler can be thought of as a digital audio memory device that lets you instantly access sound files, process them and then play them back from RAM. Once loaded, these sounds (whose length and complexity are often only limited by memory size and your imagination) can be looped, modulated, filtered and amplified (using such techniques as basic editing, looping, gain changing, reverse, music scale sample-rate pitch change and digital mixing) in a way that generally allows the results to be played in a musical and polyphonic fashion.

The Hardware Sampler

A sampler's design can include a music keyboard or set of trigger pads (Figure 5.8) that let you play samples as polyphonic musical chords, sustain pads, triggered percussion sounds or sound-effect events ... directly from its surface. It can also be designed as a sample module that integrates all of the necessary signal processing, programming and digital control structures into a rack or half-rack mountable unit.

FIGURE 5.8
Hardware samplers. Roland MC-707 Groovebox Sampler. (Courtesy of Roland Corporation U.S., www.roland.com.)

These samples can be played according to the standard Western musical scale (or any other scale, for that matter) by altering the playback sample rate over the device's note range. For example, pressing a low-pitched key on the keyboard will cause the sample to be played back at a lower sample rate, while pressing a high-pitched one will cause the sample to be played back at rates that would put Mickey Mouse to shame. By choosing the proper sample-rate ratios, these sounds can then be polyphonically played (where multiple notes are sounded at once) at pitches that correspond to standard musical chords and intervals.

A sampler (or synth) with a specific number of voices, 64 voices for example, simply means that multiple sound patches can be combined to play up to 64 simultaneous notes at any one time. Each sample in a multiple-voice system can be assigned across a performance keyboard, using a process known as splitting or mapping. In this way, a sound can be assigned to play across the performance surface of a controller over a range of notes, known as a zone (Figure 5.9). The grouping of samples into various zones and velocity splits can also enter into the equation by allowing multiple samples to be mapped across the keyboard surface or layered onto keys, allowing different samples to be played according to how soft or hard the keys are pressed. For example, a single key might be layered so that pressing the key lightly would reproduce a softly

bass gong   upright bass   soft grand piano

hard grand piano

soft honky piano

loud honky piano

recorded sample, while pressing it harder would produce a louder sample with a sharp, percussive attack. In this way, mapping can be used to create a more realistic instrument or a wild set of soundscapes that change not only with the played keys but with different velocities as well.

Most samplers have extensive edit capabilities that allow the sounds to be modified in much the same way as a synthesizer, using such modifiers as:

- Velocity
- Panning
- Expression (modulation and user control variations)
- Low-frequency oscillation (LFO)
- Attack, delay, sustain and release (ADSR) and other envelope processing parameters
- Keyboard scaling
- Aftertouch

Many sampling systems will often include such features as integrated signal processing and multiple outputs, which offer isolated channel outputs for added mixing and signal processingversatility.

FIGURE 5.9

Example of a sampler's keyboard layout that has been programmed to include zones. Notice that the upper register has been split into several zones that are triggered by varying velocities.

FIGURE 5.10

Native Instruments software samplers. (a) Kontakt Sampler; (b) battery percussion-based sampler. (Courtesy of Native Instruments GMBH, www. native-instruments.com.)

(a)

(b)

FIGURE 5.11
Simpler software
sampler for Ableton Live.
(Courtesy of Ableton AG,
www.ableton.com.)

The Software Sampler

Although hardware samplers are sought out for use by old-school producers and artists, by far the more commonly found sampling systems found in modern-day music production are software samplers (Figures 5.10–5.12). As one would expect, these software devices have improved to the point of equaling or surpassing their hardware counterparts in cost-effectiveness, power and ease of use. They offer much of the same, if not more, capabilities as their hardware counterparts, being able to edit, map and split sounds across a MIDI keyboard/controller using on-screen graphic controls with full DAW integration.

Using the digital signal processing (DSP) capabilities of today's computers (as well as the recording, sequencing, processing, mixing and signal-routing capabilities of most DAWs), these software samplers are able to store and access samples within the internal memory of a laptop or desktop computer. Using an on-screen graphic user interface, these sampling systems allow the user to:

FIGURE 5.12
Cubase/Nuendo Sampler
Track. (Courtesy of
Steinberg Media
Technologies GmbH, a
division of Yamaha
Corporation,
www.steinberg.net.)

- Import previously recorded sound files in multiple audio formats.
- Edit and loop sounds into a final usable form.
- Vary envelope parameters (e.g., dynamics) over time.
- Vary processing parameters.
- Save the edited sample setup parameters as a file for later recall.

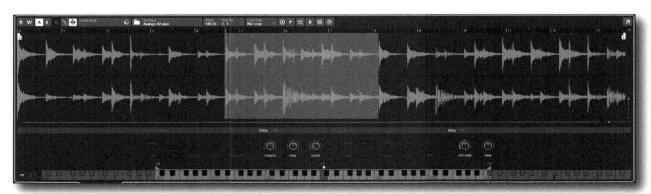

Literally, the mind boggles at the possibilities, range of styles and production tools that are offered by modern-day sampling systems. In addition to creating your own sampled sounds, a wide range of "sample packs" are available for free or for purchase from various manufacturers and on-line. The general production level ranges from amateur to pro, meaning that, whenever possible, it's wise to listen to examples to determine their quality and to hear how they might fit into your own personal or project style before you buy. As a final caveat … you've probably heard of the legal battles that have been raging over sampled passages that have been "ripped" from recordings of established artists. In the fall of 2004 decision regarding Bridgeport Music et al. vs. Dimension Films, the 6th Circuit U.S. Court of Appeals ruled that the digital sampling of a recording without a license is a violation of copyright, regardless of size or significance. This points to the need for tender loving care when lifting samples off of a CD or the Web.

THE DRUM MACHINE AND MPC-STYLE PRODUCTION

Historically, the first sample-based device ever to come onto the market in music production was the drum machine (Figure 5.13). In its most basic form, these devices make use of ROM-based, prerecorded waveform samples to reproduce high-quality drum sounds from its internal memory. These factory-loaded sounds often include a wide assortment of drum sets, percussion sets, rare and wacky percussion hits and effected drum sets… Who knows – you might even encounter scream hits by the venerable King of Soul, James Brown.

These pre-recorded samples can then be assigned to a series of playable keypads that are generally located on the machine's top face, providing a straightforward controller surface that often sports velocity and aftertouch dynamics. Each of these pads can be easily edited using control parameters such as tuning, level, output assignment and panning position.

FIGURE 5.13
SR18 Hi-Def Stereo Multi-Sampled Drum Machine. (a) Top view; (b) rear view. (Courtesy of Alesis, a registered trademark of inMusicBrands, LLC; www.alesis.com.)

(a)

(b)

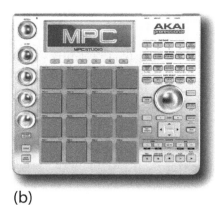

(a)   (b)

FIGURE 5.14
Akai Music Production Center
(MPC) system. (a) MPC X; (b)
MPC Studio. (Courtesy of Akai
Professional, a registered
trademark of inMusicBrands,
LLC; www.akaipro.com.)

Because of new cost-effective technology, drum machines began to include basic sampling technology, which allows sounds to be imported, edited and triggered directly from with the box. With the direct incorporation of basic sequencing capabilities, these boxes began to evolve into full beat creation systems, culminating in the Akai Music Production Center (MPC) type of system (Figure 5.14).

MPC-type systems make use of easy-to-use navigation that lets the musician record, edit, effect and sequentially perform grooves from its performance surface. Because of this ease, it was picked up by well-known hip-hop and rap artists who then spawned a whole generation of music producers.

INTEGRATED MUSIC PRODUCTION SYSTEMS

With the development of music production systems, a few companies have developed an integrated approach to offering stand-alone and plugin tools for music production. One such system is Reason from the folks at Reason Studios

FIGURE 5.15
Reason Studios music production system for recording, sequencing and mixing. (Courtesy of Reason Studios, LLC; www.reasonstudios.com.)

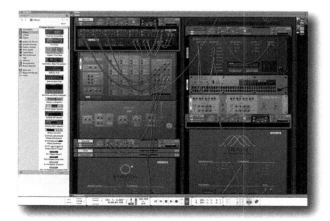

(Figure 5.15). Reason defies specific classification in that it's an overall music production environment that has many integrated facets. For example, it includes a MIDI sequencer, as well as a wide range of software instrument modules that can be played, mixed and combined within a comprehensive on-screen environment that can be controlled from any DAW or external controller.

In essence, Reason is a combination of modeled representations of vintage analog synthesis gear, mixed with digital synthesis and sampling technology. Combine these with a modular approach to signal and effects processing, add a generous amount of internal and remote mix and controller management (via an external MIDI controller), top this off with a quirky but powerful sequencer and you have software that lets you build tracks using your studio computer or laptop.

In short, Reason functions by:

- Allowing you to choose a virtual instrument (or combination of instruments)
- Calling up a programmed sample loop or an instrument performance patch
- Allowing these sounds and sequence patterns to be looped and edited in new and unique ways
- Letting you create controller tracks that can vary and process the track in new and innovative ways
- Allowing a wide range of synchronized and controllable effects to be added to the track mix

Once you've finished the outline of one track, the obvious idea is to move on to create new instrument tracks that can be combined in a traditional multitrack building-block approach, until a song begins to form. Since the loops and instruments are built from preprogrammed sounds and patches that reside on the hard disk, the actual Reason file (.rns) is often quite small, making this a perfect production media for publishing your songs on the Web or for collaborating with others around the world. Additional sounds, loops and patches are also widely available for sale or for free as "refills" that can be added to your collection to greatly expand the software's palette.

Another type of semi-integrated system is offered by the folks at Native Instruments in the form of an overall package, known as Komplete (Figures 5.16),

FIGURE 5.16
Native Instruments Komplete plugin library with 50+ virtual instruments/effects and 25,000 preset sounds. (Courtesy of Native Instruments GMBH, www.native-instruments.com.)

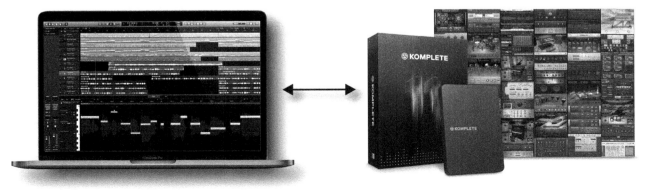

FIGURE 5.17
Komplete Kontrol controller/software integration system. (Courtesy of Native Instruments GMBH, www. native-instruments.com.)

which includes over 50 products of music production stand-alone and plugin instruments and effects, including the award-winning Kontakt 6 sampling engine, with a number of instrument libraries that can be added to by buying additional sample packs. A program block-related synth known as Reaktor is also included that allows multiple synth techniques to be built up by the user or by well-known musicians. With over 25,000 preset sounds, 10 expansion packs and 170 GB of content for the Mac/PC, Komplete is a compositional toolset that over the years has become an industry mainstay.

In order to better unify the Native Instruments product line, the folks at "NI" have created the Komplete Kontrol standard for use with NI keyboards (Figure 5.17), Maschine and other music production hardware. This protocol allows NI and other compliant software plugins to be accessible, organized and instantly mapped to the keyboard or device controls, allowing instruments, effects, presets, parameters, loops and samples to be easily accessed and played … all without the extensive need for on-screen computer control.

CHAPTER 6
Controllers

A MIDI *controller* is a device that's expressly designed to control other devices (be they for sound, light or mechanical control) within a connected MIDI system. Unlike a MIDI instrument, these devices contain no internal tone generators or sound-producing elements, but often include a high-quality performance surface and a wide range of controls for handling control, trigger and device-switching events. Since controllers have become an integral part of music production and are available in many incarnations to control and emulate many types of musical instrument types, don't be surprised to find various incarnations of these devices popping up all over this book and within electronic music production. Basically, they're a necessary tool for translating human performance and control interactions into digital data … and vice versa.

As was stated, these devices don't produce sound on their own. Instead they can be used in the studio or on the road as a simple and straightforward surface for handling MIDI performance, control and real-time switching (Figure 6.1).

For the remainder of this chapter, we'll focus on many of the controller types that are available on the market today.

KEYBOARD CONTROLLERS

With the rise of computers, sound modules, virtual software instruments and other types of digital devices have come onto the production scene that don't incorporate a music keyboard into their design, so it makes sense that the MIDI keyboard controller (Figures 6.2–6.4) has grown in popularity, to the point that it is standard in most production setups. Such an external keyboard device might easily include:

- Music keyboard surface
- Variable set of parameter controls
- Fader, mixing and transport controls
- Percussion, event or loop trigger pads
- Switching controls
- Tactile control surfaces

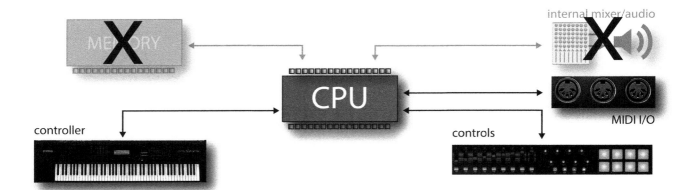

internal mixer/audio

CPU

MIDI I/O

controller

controls

FIGURE 6.1
The basic building blocks of a MIDI controller (notice that there are no sound-generating or sound-related data components).

As you might imagine, keyboard controllers vary widely in size and the number of features that are offered. For starters, the number of keys can vary from the sporty, portable 25-key models (which might offer keys that are much smaller than a full piano key size) to those having 49 and 61 keys and all the way up to the full 88-key models that can play the entire range of a full-size grand piano (often having keys that are fully or partially weighted, so as to emulate the feel of an acoustic piano's keys).

Beyond the standard pitch and modulation wheels (or similar-type controller), the number of options and general features are up to the manufacturer. This might include slider controls that allow for direct control over instrument/plugin performance parameters, while additional buttons and input controls might offer transport function control over a DAW, percussion trigger pads and the like. Others might offer on-board readouts that allow the user to directly see any parameter changes that might be made by the user or by the system under automation.

Another type of interaction involves the direct and dedicated integration between a controller and compatible software instrument or effects plugin (Figure 6.5). Direct control protocols, such as Novation's Automap or Komplete Kontrol from Native Instruments, allow the controller type, level and parameter readouts to be automatically mapped and displayed on both the controller itself and on the plugin's readout/controls, without the need for user programming. Basically, it can be thought of as a way for control parameters to be easily and seamlessly

FIGURE 6.2
Roland A-88MKII MIDI Keyboard Controller, the very first commercially available MIDI 2.0 controller. (Courtesy of Roland Corporation U.S., www.roland.com.)

mapped and displayed, giving the user simple and direct control over a wide range of parameters within plugins that conform to that particular standard.

On a completely different note … the daddy of all keyboard controllers is the MIDI grand piano (Figure 6.6). These instruments are fully functional acoustic grands that can be performed in a normal fashion, but can also output MIDI performance messages and then respond to MIDI by mechanically playing it in an automated, player-piano style. Other types, that are not true MIDI acoustic grands, might be made in a furniture style that mimics a baby mini or small grand, will actually be a synth or sample-based instrument that contains circuitry for playing back piano, percussion and other instrument types. In addition to a traditional, weighted keyboard, sustain and una corda pedals that are found on most acoustic and synth MIDI grands, these pianos can also offer lifelike control over other external keyboard instruments.

Not to be outdone, in recent times the electronic church organ has become the mother of all controllers. These mega keyboard devices have introduced MIDI into houses of worship, allowing recitals and special events to be recorded and then played back at a later time. Imagine walking down the aisle to a "Here Comes

FIGURE 6.3
Native Instruments Komplete Kontrol S88 MK2. (Courtesy of Native Instruments GMBH, www.native-instruments. com.)

FIGURE 6.4
Alesis V25 25 Key USB MIDI Keyboard. (Courtesy of Alesis, a registered trademark of inMusicBrands, LLC; www.alesis.com.)

FIGURE 6.5
Komplete Kontrol Controller/
Software Integration
System. (Courtesy of Native
Instruments GMBH, www.
native-instruments.com.)

the Bride" sequence. (Don't laugh – it's probably happening in Vegas as you read this.) Certain church organ systems have also joined the digital age by offering advanced sample playback engines and controller systems that can mimic the entire organ setup from a DAW environment … in all its surround glory.

Alternative Keyboard Controllers

A number of alternative keyboard controllers that make use of additional gesture controls have come onto the market in recent years. These devices (for the most part) have a traditional key layout, however they go beyond traditional polyphonic pressure and release controller changes, to add expressive gestural moves … such as adding pitch-bend by sliding individual fingers up-down the keyboard, or by adding continuously variable pressure response on every key (Figures 6.7 and 6.8).

FIGURE 6.6
Yamaha Enspire Disklavier
MIDI Grand Piano. (Courtesy
of Yamaha Corporation of
America, www.yamaha.com.)

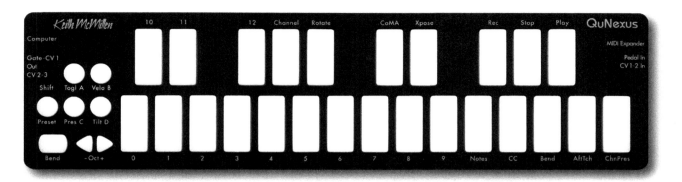

Keith McMillen

DRUM CONTROLLERS

MIDI drum controllers are used to translate the voicings and expressiveness of a drum or percussion performance into MIDI data. These "hands-on" devices are great for capturing the feel of an actual live performance, while giving you the flexibility of re-voicing, editing and automating a captured MIDI track. The form and input method of these percussion controllers can vary widely between devices, depending upon their design and the type of application that's required by the performance and production. For example:

- A standard MIDI keyboard playing surface could be used to "finger in" a beat.
- A dedicated pad controller or keyboard controller with percussion pads could be used to capture a performed beat.
- An older, dedicated drum machine (that contains its own internal sound set) and sports buttons on its top face, could be used to pound out a beat that can be recorded to a MIDI track.
- A larger set of percussion pads might allow the performer to use their hands or drumsticks to create a more realistic performance.
- A full electronic or acoustic drum set with MIDI triggers cold be used to fully capture the look and feel of an actual drum performance.

FIGURE 6.7
Keith McMillen QuNexus Keyboard. (Courtesy of Keith McMillen Instruments, www.keithmcmillen.com.)

FIGURE 6.8
ROLI Songmaker Kit Studio Edition MIDI Controller. (Courtesy of Roli, www.roli.com.)

The Keyboard as a Percussion Controller

Quite simply, the most commonly used device for triggering percussion and drum voices is the standard MIDI keyboard controller. One advantage of playing percussion sounds from a keyboard is that sounds can be triggered quickly because the playing surface is designed for fast finger movements and doesn't require full hand/wrist motion. Another advantage is its ability to express velocity over an entire range of possible values, instead of the limited number of velocity steps that are available on certain drum pad models. As percussion sounds are generally not related to any musical interval, you're also free to assign drum voices to any keyboard note and range that you'd like. In addition, percussion sounds can be assigned to a particular range of notes over a split keyboard arrangement, allowing various sound patches or percussion sounds to be layered over the playing surface.

Pad Controllers

One of the most straightforward of all drum controllers is the 4 × 4 or 8 × 8 trigger-pad design that is now built into many drum machines, portable percussion controllers and controller pad styles (Figures 6.9 and 6.10). By calling up the desired setup and voice parameters, these small footprint triggers let you go about the business of using your fingers to do the walking through a performance or sequenced track. It's also a simple matter to trigger other devices and sounds from these keypads. For example, you could assign the pads to a channel that's being responded to by your favorite synth, sampler or groove machine plugin and then trigger these sounds directly. Obviously, coming under the "Don't try this at home" category, these controller pads are generally too small and not durable enough to withstand drumsticks or mallets. For this obvious reason, they're generally played with the fingers.

Drum Pad Controllers

FIGURE 6.9
Korg nanoPAD2 Pad Controller. (Courtesy of Korg Inc., www.korg.com.)

In more advanced MIDI project studios or live stage rigs, it's often necessary for a percussionist to have access to a playing surface that can be played more like a real instrument. In these situations, a dedicated drum pad controller

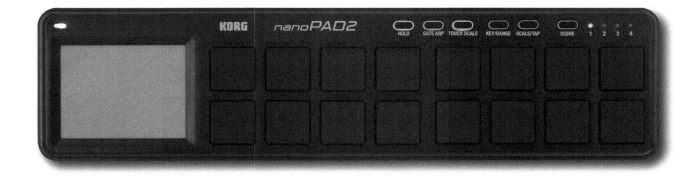

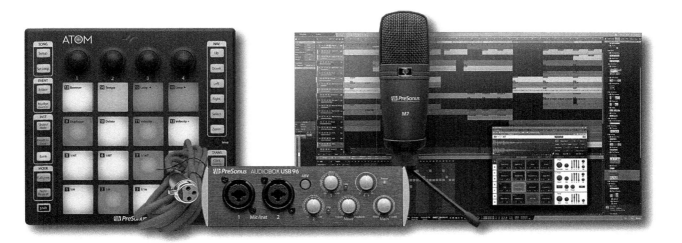

(Figure 6.11) would be better for the job. Drum pad controllers vary widely in design. They can be built into a single, semi-portable case, often having between six and eight playing pads, or the triggers can be individual pads that can be fitted onto a special rack, traditional drum floor stand or fitted into a drum set. Of course, these pads can be triggered over the full velocity range (allowing for sample splits to be built up), and their playing surface can be played with the fingers, hands, percussion mallets or drumsticks. Besides giving us a more realistic performance interface and full MIDI capabilities, one of the biggest side advantages of a drum pad setup is silence. You could bang away at a full set of pads in a New York high-rise at 3 a.m., and nobody'd ever know. Often, they're small enough that they can be conveniently tucked away in a corner of a bedroom or mid-sized project studio.

MIDI Drums

Another way to play drums without making your neighbors hate you, while also reducing your physical footprint in the studio, is to spring for a MIDI drum set

FIGURE 6.10
Atom Producer Lab, including Atom pad controller/plugin software. (Courtesy of Presonus Audio Electronics, www.presonus.com.)

FIGURE 6.11
Drum/Percussion pad controllers. (a) Alesis SamplePad Pro Percussion Pad controller. (Courtesy of Alesis, a registered trademark of inMusicBrands, LLC; www.alesis.com.) (b) Roland Octapad SPD-30 Percussion Controller (black). (Courtesy of Roland Corporation U.S., www.roland.com.)

(a)

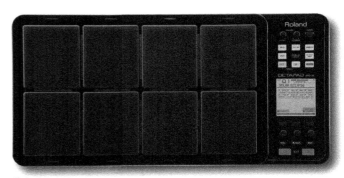

(b)

ALESIS

FIGURE 6.12
Alesis DM10 MKII Pro
Electronic Drum Set.
(Courtesy of Alesis, a
registered trademark of
inMusicBrands, LLC;
www.alesis.com.)

(Figure 6.12). These setups mimic the drums in ways that allow the performer to move naturally, while capturing the performance directly to a MIDI track. In addition to a natural drum performance, these setups offer the advantages of complete control over the chosen playback samples, expression, tuning and more, all with the advantages that come with offering complete instrument isolation.

MIDI-fying Acoustic Drums

Taking realism even further, acoustic drums can be MIDI-fied in a number of interesting ways without too much fuss. As was previously said, one of the simpler of these is to incorporate MIDI drum pads into a standard acoustic drum set. Such a setup gives us the power and sound of a traditional kit, with the added versatility, unique sounds and sequence capabilities that MIDI has to offer.

Another way to MIDI-fy an acoustic drum is through the use of trigger technology. By placing removable transducers called triggers onto the body of a drum near the heads, it's possible to make use of its pickup (often a piezo-electric transducer) to translate a drum's sound into an electrical signal that can be sent to a trigger interface (Figure 6.13). Here, the trigger signals can be translated into MIDI notes that can be recorded to a MIDI track for further sample layering, editing and production … either on stage or in the studio. Of course, specially designed pads can also be incorporated directly into a kit, allowing sounds to be triggered during a performance.

Replacing Drum Tracks

Although it doesn't make use of MIDI, it's worth mentioning that readily available software can take existing individual drum tracks (kick, snare, etc.) that were poorly recorded or are simply of the wrong timbre and automatically replace them with instruments that better fit the needs of the project (Figure 6.14). Since the timing and dynamics of the original track are used as the source for the

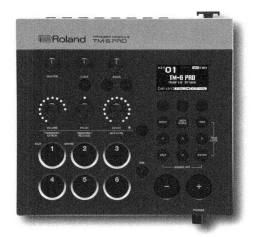

replacement sample, all of the musical intentions of the live player or sequenced track will be precisely matched.

Alternative Percussion Devices

Unlike drum machine pads or keyboard controllers, the MIDI vibraphone is generally used by professional percussionists who want a traditional playing surface while making use of the power of MIDI. These vibes are commonly designed with a playing surface that can be fully configured using its internal setup memory to provide for user-defined program changes, playing-surface splits, velocity, after touch, modulation, etc.

Just as it's possible to use your hands, mallets, sticks or whatever to humanly perform a part to MIDI using a percussion control surface … it's also possible to go the other way. Through the use of purchased or home-made devices that make use of solenoids to mechanically strike a surface, a MIDI sequence can be played out to a series of robotic actuators that can play a percussion instrument (Figure 6.15). For example, a sequence can be used to control a small

FIGURE 6.13
By using a MIDI trigger device (such as the Roland TM-6 PRO Trigger Module – shown top and back), a drum's acoustic signal can be picked up and converted to MIDI for sequencing. (Courtesy of Roland Corporation, www.roland.com.)

FIGURE 6.14
Steven Slate Drums Trigger 2 Drum Replacement Plug-in. (Courtesy of Steven Slate Drums, www.stevenslatedrums.com.)

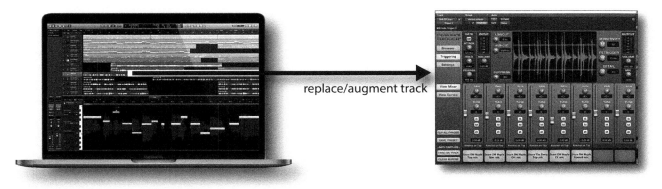

replace/augment track

FIGURE 6.15
Polyend Perc Drumming
Machine. (Courtesy of
Polyend, www.polyend.com.)

mechanical device that can be used to play out a beat on a snare. In this way, the instrument could be miked and placed into the project, allowing the drum set and studio acoustics to fully come through.

MAKING BEATS MPC STYLE

This section is a percussive "horse of a different color" in a way that has literally formed and evolved the musical, artistic and personal identity of youth in modern western culture. I'm talking about the MPC type of beat production that originally sprang onto the scene in 1988 in the form from the Akai Music Production Center.

FIGURE 6.16
The Akai Music Production
Center (MPC) system
can be used stand-alone
or in conjunction with a
DAW. (Courtesy of Akai
Professional, a registered
trademark of inMusicBrands,
LLC; www.akaipro.com.)

Used by iconic rap, hip-hop and countless other modern genres, this small device that was originally slightly bigger than an Etch-a-Sketch went well beyond the capabilities of a drum machine. Like its earlier incarnation, the MPC (Figure 6.16) did contain "outta-the-box" sampled sounds that could be easily triggered from any of its 16 top-facing pressure-sensitive pads. It, however, also

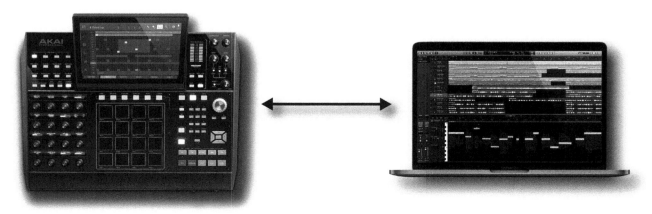

contained an on-board sequencer and digital system that allowed the user to sample their own beats (often from a commercial vinyl), edit them into smaller measures or individual "hits" that could then be assigned to one of the pads and then manually perform these samples into a new and unique groove. Its simplicity and overall production power made it easily accessible to the music production community, allowing it to be used by countless pros as their production tool of choice for creating beats, as well as full productions.

Over the decades, these big, familiar pads have been imitated by countless other hard- and software companies, which have likewise combined MPC-like production control surfaces with software recording, editing and compositional systems that are all-in-one beat production machines.

This brings us to one of the most popular incarnations within the MPC production-styled environment ... Maschine from Native Instruments (the Deutsch word for machine is feminine, therefore the ending "e" is properly pronounced as "mashin-eh"). This family of devices uses a USB controller/software combination (Figure 6.17) to create an MPC-like production environment that lets you perform, edit and produce beats in a track-styled on-screen environment that can then be layered and combined as a stand-alone production center, or can easily integrate (as an instrument plugin) with a DAW to create finished music productions. Later incarnations of Maschine also include a full-blown, high-quality audio interface that can be used to power your entire audio production facility.

Users can assign drum kits, instruments and sounds that come packaged with the device to any of the controller's 16 pads. These sounds can be used out-of-the-box or be user-edited to create new and interesting sounds. Additional, thematic sound expansion packs can be purchased to add to your sonic library; however, in true MPC style, you can record, import and manipulate your own sounds to create your own arsenal of personal beats, hits and grooves.

FIGURE 6.17

Native Instruments Maschine MK3 Production and Performance System. a. Hardware controller. b. Stand-alone or plugin software. (Courtesy of Native Instruments GMBH, www.native-instruments.com.)

(a)

(b)

One of the best ways to get an overall sense of the world of beat-making through the lens of the MPC or other styled beat production units is to take a visual stroll through the limitless number of YouTube videos that exist on the subject. You might start with your favorite artist and see if they've created a video that might explain their personal approach to beat making. It's always good to see how others approach their personal art-making process through the use of technology.

Rolling Your Own

It's nice to open up a box, plug it in and get started right away with stock sounds and beats … always the best way to get to know a new toy. Beyond this, a virtually unlimited number of sound libraries can also be found on the Web that might be already edited and formatted for your specific sampling production tool (Figure 6.18). These libraries will almost always include a never-ending number of percussion instruments and drum sets that can be loaded into a DAW to build up MIDI drum and rhythm tracks, or you can lift complete audio percussion loops from loop sources and libraries that are readily available on the Web. One thing that I'll often do is manually re-record the sounds and loops from older, vintage groove machines and even music toys, edit them and then work that into a production in new and interesting ways … it's a trick that I have in my personal music loop library.

Beyond this, however, it's also a lot of fun and important for creative individualism to take new and original material and make a project your own by recording, creating and editing your own sounds. This can be done in a never-ending number of ways, by taking a portable hand recorder on your travels; you could pull it out and record the striking of a cabin mailbox in the hills of Austria, the clanging of a gamelan set of gongs in Indonesia or the notes of an old marimba in Mexico.

FIGURE 6.18
Several of the many available sample packs for the Maschine Production and Performance System. (Courtesy of Native Instruments GMBH, www. native-instruments.com.)

If you can't find the sounds you want, then create your own! It'll give you a great chance to personalize your music and/or effects, in a way that will often cause everyone's ears to perk up. As always, the sky's only limited by your imagination.

GUITAR AND BASS GUITAR CONTROLLERS

Guitar players of all types often work at stretching the vocabulary of their instruments beyond the norm. They love doing non-traditional gymnastics using such tools of the trade as distortion, phasing, echo, feedback, etc. Due to advances in guitar pickup and microprocessor technology, it's now possible for the notes and minute inflections of guitar strings to be accurately translated into MIDI data (Figure 6.19). With this innovation, many of the capabilities that MIDI has to offer are now available to the electric (and electronic) guitarist. For example, a guitar's natural sound can be layered with a synth pad that's been transposed down to give a rich, thick sound that'll shake your boots. Alternatively, recording a sequenced (and probably live) guitar track into a project could give a producer the option of changing and shaping the sound in mixdown. On-stage program changes can also be a big plus for the performing MIDI guitarist. This on-stage tool lets the player switch between guitar voices, synth voices and anything in between by simply stomping on a MIDI foot controller.

FIGURE 6.19
G-5 VG MIDI Stratocaster with Roland GR-55 guitar synth. (Courtesy of Roland Corporation U.S., www.roland.com.)

FIGURE 6.20
Yamaha WX5 Wind MIDI Controller. (Courtesy of Yamaha Corporation of America, www.yamaha.com.)

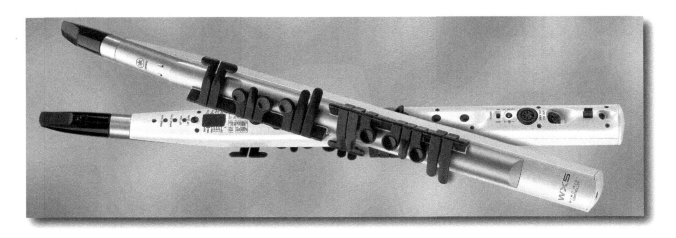

In addition to having a physical system that can capture the nuances of the actual guitar strings, software also exists that can translate the audio track of a live or recorded electric or acoustic guitar directly into low-latency, polyphonic MIDI messages. In this way, a performance or previously recorded track could be converted to MIDI for further production.

WIND CONTROLLERS

MIDI wind controllers (Figure 6.20) are expressly designed to bring the breath and key articulation of a woodwind or brass instrument into the world of MIDI performance. These controller types are used because many of the dynamic- and pitch-related expressions (such as breath and controlled pitch glide) simply can't be communicated from a standard musical keyboard. In these situations, wind controllers often help create a dynamic feel that's more in keeping with their acoustic counterparts by using an interface that provides special touch-sensitive keys, glide- and pitch-slider controls and sensors for outputting real-time breath control over dynamics. It's interesting to note that during the recording of Daft Punk's *Random Access Memories* "The Game of Love", they not only recorded the MIDI from the wind controller part, they actually recorded the performer's breath as well, to give realism to the track.

FIGURE 6.21
Performance pad controllers.
(a) Novation Launchpad MkII.
(Courtesy of Novation Music, a brand of Focusrite Audio Engineering Plc, www. novationmusic.com.)
(b) Ableton Push.

PERFORMANCE PAD CONTROLLERS

Another class of controllers differs from its instrument-like counterparts, by making use of a playing surface that physically mimics its acoustic equivalent. These hard- or software performance pad controllers (Figures 6.21 and 6.22)

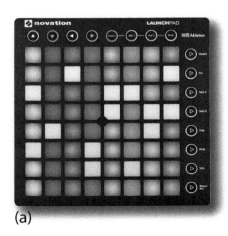

(a)

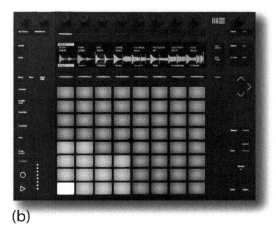

(b)

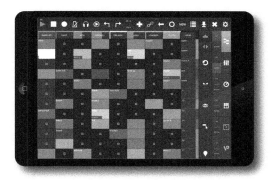 

will generally offer up a grid of buttons that can be used to trigger an instrument, or more likely, a sample and loop-segment within such a DAW as Ableton Live. When used with these programs in a live setting, individual clips and/or scenes can be launched to create a fully interactive stage performance.

These wired or wireless controllers can also be used in the studio or on-stage to switch between controller functions, such as sound file/loop triggering, extensive parameter control or on-screen mixing … all from a single surface. I have personally used these controllers to perform on-stage, record that performance and then re-integrate the stage performance back into a production session, so as to create a live-event version of a project … and I got several Grammy nominations for them. Having fun and going bold can definitely pay off!

FIGURE 6.22

Touchable 3 can be used to wirelessly control Ableton Live in a practice and performance setting. (Courtesy of Zerodebug, www.touch-able.net.)

CHAPTER 7

Groove Tools

The expression "getting into the groove" of a piece of music often refers to a feeling that's derived from the underlying foundation of music – rhythm. With the introduction and maturation of MIDI and digital audio, new and wondrous tools have made their way into the mainstream of music production. These tools that can help us to use technology to forge, fold, mutilate and create compositions that make direct use of rhythm and other building blocks of music through the use of looping technology. Of course, the cyclic nature of loops can be overly repeat-repeat-repetitive in nature, but new toys and techniques for looping have expanded the flexibility, control, real-time processing and mixing in ways that can be used by an artist in wondrously expressive ways.

In this chapter we'll be touching upon many of the approaches and software programs that have evolved (and continue to evolve) into what is one of the fastest, interactive and most accessible facets of personal music production. It's literally impossible to hit on all of the finer operational points for these systems; for that, I'll rely upon your motivation and drive to:

- Download many of the software demos that are readily available
- Delve into their manuals and working tutorials, then play to your heart's content
- Begin to create your own grooves and songs that can then be integrated with your own music or those of collaborators

If you do these three things, you'll be surprised as to how much you've learned in ways that can directly translate into skills that'll widen your production horizons and most likely impact your music.

THE BASICS

Because groove-based tools often deal with rhythms and cyclic-based loops (Figure 7.1), there are several factors that need to be managed in order to be musically and technically effective:

- Sync
- Tempo and length
- Pitch

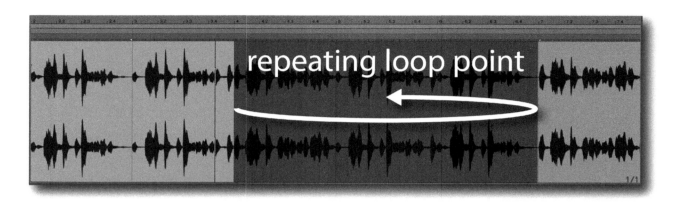

repeating loop point

FIGURE 7.1
Grooves or rhythm loops will most often be cyclic in nature.

The aspect of sync relates to the fact that the various loops in a groove project will need to line up in time with each other (or in multiple lengths and timing intervals of each other). It almost goes without saying that multiple loops that are successively or simultaneously triggered must have an exact timing relationship to one another – otherwise, it's a non-musical, jumbled mess.

The next relationship relates to the aspect of tempo. Just as it's critical that synchronization between the loops be kept in strict lockstep with each other, it's also an important part of the creative process that each soundloop be adjustable in pitch (resampled) and/or in length (time stretched). This allows for complete production flexibility, letting the overall session tempo and individual loop pitch and length to be varied to fit the production needs of the song or program content.

Time and Pitch Change Techniques

The process of altering a sound file to match the current session tempo and to synchronously align them occurs in the software by combining variable sample rates with pitch-shifting techniques that can be applied to each loop file within the session. Using these basic digital signal processing (DSP) tools, it's possible to alter a sound file's duration (by raising or lowering its playback sample rate to alter the program length) or to alter its relative pitch (either up or down). In this way, three possible combinations of time and pitch change can occur:

- Time change: allowing the loop file length (playback speed) to be altered without affecting its pitch.
- Pitch change: allowing the loop file pitch to be altered, while the length remains unaltered.
- Both: allowing both the loops pitch and length to be altered using basic resampling techniques.

Most commercially available loop-based programs and plugins will usually involve the use of recorded sound files that are encoded with special headers (a form of metadata) that include information on their original tempo and length (in both samples and beats). By setting the loop program to a master tempo (or a specific tempo at that point in the song), an audio segment or file can be imported, examined as to its sample rate or length … and then be recalculated to a desired pitch and relative tempo that matches the session's user-programmed variables. Voila! We now have a defined segment of audio that can fit within the session tempo, allowing it to play and interact in relative sync with the other defined loop files.

WARPING

This process of altering the sample rate to vary the speed (tempo) and pitch (key) of a loop file will often make use of a process called *warping*. This technique uses the original loop's file tempo and sample rate to determine key variables that can be used to alter its parameters. These algorithms, however, may go further by using complex detection techniques to detect the beats and other percussive transient cues to create "hit points". The use of hit points (Figure 7.2) essentially allows a detection algorithm to create a dynamic tempo map of the loop file's rhythmic structure, in such a way that variations in the beat can actually be smoothed out or altered to match the original tempo. Once a hit point has been detected and entered, it can be automatically or manually moved in time to a place that matches the timing division marks of another track. By moving this defined point, the software is able to speed or slow the file's playback as it moves from one hit point to the next, allowing the relative timing differences to be smoothed out … or they can be manually edited by the artist to create a whole new set of metric time divisions and variations that can have a subtle or not-so-subtle effect upon a song.

Before continuing on, I feel that it's important to point out that pretty much all apps, programs and DAWs that are capable of dealing with loops will program their own set of warping algorithms. Often a number of pre-programmed algorithm choices will be offered, depending on if the loop file is percussion/drum-based, melodic in nature or is based upon chords. Choosing a percussion algorithm on a loop that contains only a melody might sound really bad (and vice versa). In addition, many high-end DAWS will offer a wide range of choices for interpolating (a fancy word for guessing) the original sound file and determining the best time/pitch shift parameters. It's important to note that the

FIGURE 7.2
Hit point markers can be used to show and manipulate events and percussive transients in a sound file. (Courtesy of Ableton AG, www.ableton.com.)

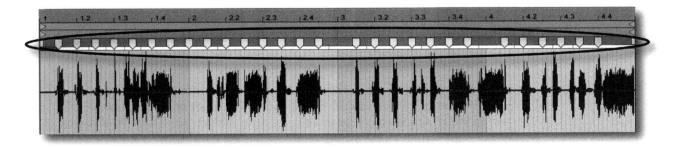

time/pitch shift algorithms in one program will often sound radically different to those of another program.

In essence, the creation of hit points by a DAW will have two purposes:

1. To detect the transients (most often with drum or percussive samples), interpolate (guess) where the hit points will fall and then smooth out the timing variations, so that the all-important beat is more "in-time".

2. To detect the transients, interpolate the hit point marker placements and then allow the artist to move the hit points, so that they can create an entirely new set of rhythmic variables … in short, to mix things up a bit … or a lot.

Once you've taken the time to better understand how your DAW or other software apps deal with hit points, experimentation, personal taste and your ears will be your best guide. This process is, in the end, definitely a musical artform.

BEAT SLICING

As we read, most loop-based production tools make use of time-stretch, pitch-shift and warping techniques to combine any number of random loop files into a session that is cohesive in both pitch and time. Another older, but still used, method called *beat slicing* makes use of an entirely different process to match the length and timings of a sound file segment to the session tempo. Rather than changing the speed and pitch of a sound file, the beat slicing process actually breaks an audio file into a number of small transient-defined segments (not surprisingly called slices) and then changes the length and timing elements of a segment by adding or subtracting the time that falls between these slices (Figure 7.3). This radically different procedure has the advantage of preserving the original pitch, timbre and sound quality of a file, but has the potential downside of creating silent gaps in the final playback (which are usually masked by the other beats and music in the song).

The most universally used format for beat slicing is the REX file, which was originated by Propellerhead. These files can be found in countless sample libraries as pre-formatted beat slice loops that can simply be loaded into most of the currently available DAWs (although you may need to install Propellerhead's free Rex Shared Library in order to import and work with these loops). If you would like to edit and create your own loops from your own sound files, Propellerhead's ReCycle program can be used.

In short, REX files don't change the pitch, length or sound quality of the loop, in order to match the current DAW's session tempo. Instead, these transient detected hit-point slices are compressed or spread out in time to match the current tempo. When the tempo is increased, the slices are pulled together in time and simply overlap. When the tempo is slowed, the slices are spread out in time by inserting just the right amount of silence. Too slow of a tempo will actually cause audible

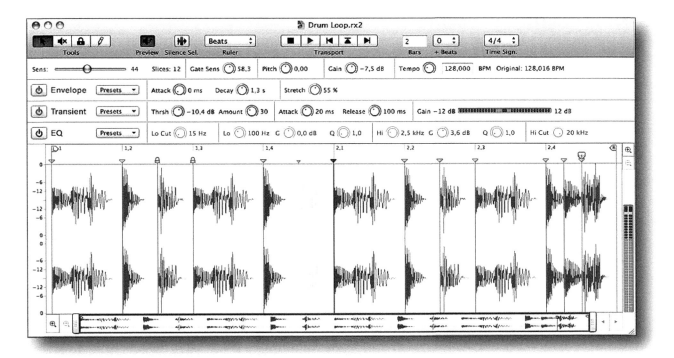

gaps to fall between the cracks. In order to reduce this effect, a "stretch" algorithm can be used that doesn't actually time-stretch the loop file, but instead uses a small sustain section within the loop, reverses it and adds it to the end of the file, essentially smoothing out the gaps with quite acceptable results.

From this, it's easy to understand why the beat-slicing process often works best on percussive sounds that have silence between their individual hit points. This process is usually carried out by detecting the transient events within the loop and then automatically placing the slices at their appropriate points (often according to user-definable sensitivity and detection controls).

As with so many other things we've discussed, the sky (and your imagination) is the limit when it comes to the tricks and techniques that can be used to match the tempos and various time-/pitch-shift elements between sliced loops, MIDI and sound file segments. I strongly urge you to take the time to read the software manuals to learn more about the actual terms and techniques, so you can best put them into practice.

FIGURE 7.3
ReCycle Groove Editing Software. (Courtesy of Reason Studios Software, www.reasonstudios.com.)

LOOPING YOUR DAW

Most DAWs offer features that can make it possible to incorporate looping into a session, alongside of other audio, MIDI and video files. Even when features that could make a specific task much easier aren't available, it's often possible

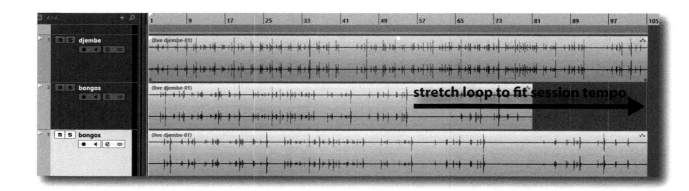

FIGURE 7.4

Commonly, a loop or sound file can be stretched to match the overall session tempo.

to think through the situation and find workarounds that can help tackle the problem at hand. For example, by manually placing the beginning point of an edited segment at the beginning of a measure, it's usually easy to stretch a segment (use pitch and time to alter the length), so that its length matches the proper session tempo. Said another way, if a session has been set to a tempo of 94 bpm, an 88-bpm loop can be imported at a specific measure. Then, by turning on the DAW's snap-to-grid and automatic time-stretch functions, the segment can be manually time-stretched until it snuggly fits into the session's native tempo (Figure 7.4). Now that the loop fits, it can be manually looped (i.e., copied) to your heart's content. Just remember, there are often no hard and fast rules. With planning, ingenuity and your manual's help, you'll be surprised at the number of ways that a looping problem can be solved and turned into an opportunity.

DIY
do it yourself

DIY: MANUAL LOOPING WITH A DAW

- Search the Web for "royalty-free loops" and download a favorite sample pack set of loops.
- Load the individual loops into your favorite DAW.
- Set the time display to read in bars/beats and set the session's tempo (in bpm) to the sample pack's native tempo and import the loops into the session.

- Copy/paste the loops into a fun song of your own and save the session.
- Import or create a loop of any tempo into a new track and use the DAW's stretch feature (consult your manual) so that it also fits the session tempo.
- Play around with all of the loops and have fun!

GROOVE AND LOOP-BASED PLUG-INS

Of course, it's a sure bet that for every hardware looping tool, there are software plugin groove tools and toys (often with their associated hardware controllers) that can be inserted into your DAW. These amazing software wonders often make life easier by:

- Automatically following the session tempo
- Allowing internal I/O routing to the DAW's mixer
- Making use of the DAW's automation and external controller capabilities
- Allowing individual groove sub-parts or combined groove loops to be imported into a session

These software instruments come in a wide range of sounds and applications that can often be edited and effected using its on-screen user interface, which can usually be remotely controlled from an external MIDI controller.

Loop-based audio editors are groove-driven music programs (Figure 7.5) that are designed to let you drag and drop pre-recorded or user-created loops and audio tracks into a graphic multitrack production interface. At their basic level, these programs differ conceptually from their traditional DAW counterpart in that the pitch- and time-shift architectures are so variable and dynamic that, even after the basic rhythmic, percussive and melodic grooves have been created, their tempo, track patterns, pitch, session key, etc. can be quickly and easily changed at any time. With the help of custom, royalty-free loops (available from many manufacturer and third-party companies), users can quickly and easily experiment with creating grooves, backing tracks or creating a sonic scape by simply dragging the loops into the program's main sound file view … where they can be arranged, edited, processed, saved and exported.

One of the most interesting aspects of the loop-based editor is its ability to match the tempo of a specially programmed loop to the tempo of the current session. Amazingly enough, this process isn't that difficult to perform, as the program extracts the length, native tempo and pitch information from the imported file's header and

FIGURE 7.5
Apple GarageBand™.
(Courtesy of Apple
Computers, Inc.;
www.apple.com.)

(using various digital time- and pitch-change techniques) adjusts the loop to fit the native time and pitch parameters of the current session. This means that loops of various tempos and musical keys can be easily and automatically adjusted in length and pitch so as to fit in time with previously existing loops … just drag, drop and go!

Of course, the graphic user interfaces (GUIs) between looping software editors and tools can differ to a certain degree. Most layouts use a track-based system that lets you enter or drag a preprogrammed loop file into a track and then drag (or copy) it to the right in a way that repeats the loop in a traditional way. Again, it's worth stressing that DAW editors will often include such looping functions that can be basic (requiring manual editing or sound file processing) or advanced in nature (having any number of automated loop functions).

Other loop programs make use of visual objects that can be triggered and combined into a real-time performance and mixing environment by clicking on either an icon or track-based grid. One such program is Live from Ableton Software (Figure 7.6). Live is an interactive loop-based program that's capable of recalculating the time, pitch and tempo structure of a sound file or easily defined segment, and then entering that loop into the session at a defined global tempo. Basically, this means that segments of any length or tempo that have been pulled into the session grid will be recalculated to the master tempo and can then be combined, mixed and processed in perfect sync with all other loops in the project session.

In Live's Arrangement View, as in all traditional sequencing programs, everything happens along a fixed song timeline. The program's Session View, however, breaks this limiting paradigm by allowing sound or MIDI files to be mapped onto a grid as buttons (called clips). Any clip can be triggered at any time and in any order in a random fashion that lends itself amazingly well to interactive performance both in the studio and on-stage (as a live performer, I'm well aware of its amazing capabilities). Functionally, each vertical column or track can play only one clip at a time. Horizontally, any number of clips can be played in a loop fashion by clicking on their respective launch buttons. By clicking on the

FIGURE 7.6
Ableton Live performance audio workstation: (a) Arrangement View; (b) Session View. (Courtesy of Ableton AG, www.ableton.com.)

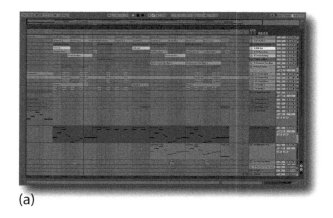
(a)

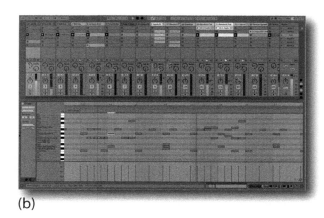
(b)

scene launch at the screen's right, every clip in a row will simultaneously play. Basically, the idea is to drag loops into the clip buttons on the various tracks, add effects and control parameters and you're well on your way to making a production. All you need to do to have fun is to start clicking.

Of course, this and other looping software packages are capable of incorporating MIDI and MIDI instruments into a project or live interactive performance. This can be done through the creation of MIDI sequence patterns or drum pattern clips that allow performance data to be built up as a series of loop clip patterns.

DIY: HAVING FUN WITH A LOOP-BASED EDITOR

- Download and load a demo of your favorite loop-based editor.
- Search the Web for "royalty-free loops" and download a favorite sample pack or choose from any number of loops that might have installed with the demo.
- Preview any number of loops and begin to drag them into the program.
- Read the program's manual (if you feel the need) and begin to experiment with the loops.

- Mess around with the tempo and various global/individual musical key signatures.
- Copy, duplicate and/or trigger the various loop tracks to your heart's content until you've made a really great song!
- Save the session to play for and share with your friends!

FL Studio

FL Studio (formerly known as Fruity Loops) is a full-featured melody and loop creation software package that includes a wide range of virtual instruments, MIDI and audio loops, as well as MIDI and audio effects. This stand-alone piece of software (Figure 7.7) allow for loops and/or MIDI segments to be placed and easily looped, edited and composed into its production window screen. A full-featured mixer is also integrated into the system, thereby offering an all-in-one production environment that's being used by a large number of music producers and artists.

Reason

Another software-based groove tool that makes use of an entirely different mind-set is Reason from the folks at Reason Studios (Figure 7.8). Reason defies specific classification in that it's an integrated music production environment includes as a wide range of included software instruments (synths, samplers, etc.) and effects, as well as its own digital audio recording capabilities, MIDI sequencer

FIGURE 7.7
FL Studio MIDI and Loop-base production workstation. (Courtesy of Image Line Software, www. image-line.com/flstudio.)

FIGURE 7.8
Reason music production software. (Courtesy of Reason Studios, www.reasonstudios.com.)

and mixer. This production package can be used as a stand-alone program that lets you create a music production from start to finish, or it can be used (via ReWire) in conjunction with your favorite DAW.

In essence, Reason is a combination of modeled representations of vintage analog synthesis gear, mixed with the latest in digital synthesis and sampling technology (Figure 7.9). Combine these with an old-school, yet high-tech software approach to signal and effects processing, add a generous amount of internal and remote mix and controller management (via an external MIDI controller) and then top this off with a quirky but powerful sequencer, and you have a software package that's powerful enough for top-flight productions using your laptop in your seat on a crowded plane. When asked to explain Reason to others I'm often at a loss, as the basic structure is so open-ended and flexible that the program can be approached in as many ways as there are people who

(a)

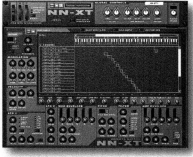

(b)

produce on it. That's not to say that Reason doesn't have a signature sound – it often does; however, it's an amazing tool that can be either used on its own or in combination with other production instruments and tools.

Of course, once you've finished the basic building blocks of a song, the idea would be to keep adding instrument tracks to the production, so as to build it up (using a traditional track-based approach to sequencing). Since the loops and instruments are built up from preprogrammed sounds and patches that reside on hard disk, the actual Reason file (.rns) is often quite small, making this a popular format for publishing your songs on the Web or for collaborating with others around the world. Additional sounds, loops and patches are widely available for sale or for free as "refills" that can be added to expand the software's production palette.

FIGURE 7.9

Reasons instruments examples. (a) SubTractor polyphonic synth module; (b) NN-XT sampler module. (Courtesy of Reason Studios, www.reasonstudios.com.)

Loop-Based Hardware/Software Systems

As was stated in the chapter on controllers, there are a number of very popular groove systems that combine a software plugin (or stand-alone app) with a connected hardware counterpart. These systems (Figure 7.10), which are often

FIGURE 7.10

Groove-based software/controller systems: (a) Maschine groove hardware/software production system. (Courtesy of Native Instruments GmbH, www.nativeinstruments.com.) (b) Spark Drum Machine. (Courtesy of Arturia, www.arturia.com.)

(a)

(b)

percussive in nature, can be used to produce music and grooves in a wide number of ways. Systems such as Maschine from Native Instruments allow us to work in an MPC-like environment, while Spark from Aurturia takes more of a drum machine/pattern creation approach to working with loops.

REWIRING GROOVE TOOLS

As a quick note, in Chapter 8, we will learn that ReWire and ReWire2 are special protocols that were developed by Reason Studios and Steinberg to allow audio to be streamed between two, simultaneously running computer applications. Unlike a plugin, where a task-specific application is inserted into a compatible host program, ReWire allows the audio and timing elements of a supporting program to be seamlessly integrated into another host program that supports the protocol.

The virtual instruments and or groove tools within most rewire-able software programs offer up advantages for gaining direct access to the MIDI, performance and mixer paths of a DAW. In this way, one or more virtual instruments and production tools will have direct I/O access to your DAW. Once you've finished laying down your grooves and the MIDI data have been saved to their respective tracks, the ReWired instruments could then be exported (bounced) to an audio tracks within the DAW and … you're off and running.

GROOVE AND LOOP HARDWARE

Truth be known, the software world doesn't actually hold the total patent on looping tools and toys, there are a number of hardware groove keyboards and module boxes that can be found on the new and used markets. These systems, which range widely in sounds, functionality and price, can offer up a vast range of unique sounds that can be quite useful for laying a foundation under your latest production.

In the past, getting a hardware grove tool to sync into a session could be time-consuming, frustrating and often problematic, requiring time and tons of manual reading; however, with the advent of powerful time- and pitch-shift processing within most DAWs, the sounds from these hardware devices can sometimes be pulled into a session without too much trouble. This brings me to offer up two, possible suggestions that might make the process of syncing sounds that are generated by hardware grooveboxes easy (or easier) within a music project session:

- As was mentioned earlier in this chapter, a single groove loop (or multiple loop) sound file could be recorded from the hardware instrument into a DAW at a bpm that's near to the session's native tempo. This can then be edited to smooth out any problems at its beginning and end and finally, imported into the session, where it can be moved to the beginning of a measure. By placing the DAW into its time stretch mode (in a way that keeps the pitch unchanged), the loop's end can be dragged so as to snap to the end of the bar. Your loop should now be in sync and can be copied within the production to your heart's content.

- The next way can be found within Chapter 11 (sync), whereby the MIDI sync on the groove hardware will be switched to "external sync". You'll then want to go to your DAW and turn on the "transmit MIDI sync". Once you have your ports and instruments configured, hitting play on the DAW should now instruct the instrument to begin playing the selected sounds in sync with the project. As the MIDI sync timing will be derived by the DAW and not the hardware instrument, you can then route the instrument audio outs to the inputs on your DAW and you're ready to record.

PULLING LOOPS INTO A DAW SESSION

When dealing with loops in modern-day production, one concept that needs to be discussed is that of importing loops into a DAW session. As we've seen, it's certainly possible to use the ReWire application to run a supporting client program in conjunction with the host program or even to use an instrument or groove-based plugin within a compatible session. However, it should be noted that often (but not always) some sort of setup may be required in order to properly configure the application. Or at worst, these applications might use up valuable DSP resources that could otherwise be used for effects and signal processing.

One of the best options for freeing up these resources is to export the instrument or groove to a new audio track. This can be done in several ways (although with forethought, you might be able to come up with new ways of your own):

- The instrument track can be soloed and exported to an audio track in a contiguous fashion from the beginning "00:00:00:00" of the session to the end of the instrument's performance.
- In the case of a repetitive loop, a defined segment (often of a precise length of 4, 8 or more bars that occurs on the metric boundaries) can be selected for export as a sound file loop. Once exported to file, it can be imported back into the session and looped into the session.
- In the case of an instrument that has multiple parts or voices, each part can be soloed and exported to its own track – thus giving a greater degree of control over individual voicings during a mixdown session.

Collecting Loop Files

In this day and age, many DAWs and groove/looping software will come packaged with their own loops or third-party, free-use loop bundles. However, there are additional ways that you can collect, create and alter loop sets for use in your projects … these include:

- Free loops and bundles that can easily be found on the Web

- Commercial loops and bundles
- Free demo and magazine content files that can be found on CD, DVD or the Web
- Rolling your own (creating your own loops can add a satisfying and personal touch to your collection)

As with most music technologies, the field of looping and laying tracks into a groove continues to advance and evolve at a fast pace. It's almost a sure bet that your current system will support looping in one way or another. Take the time to read the manuals, gather up some loops that fit your own personal style and particular interests and start working them into a session. It might take time to master the art of looping – and then again you might be a natural Zen master. Either way, it's a journey that's never-ending, tons of fun and could help you with your musical craft.

CHAPTER 8

Audio and the DAW

Besides the coveted place of honor in which most electronic musicians hold their instruments, the most important device in a MIDI system is undoubtedly the personal computer. Through the use of software programs and peripheral hardware, the desktop or laptop computer is often used to control, process and distribute information relating to music performance and production from a centralized, integrated control position.

Of course, we're all aware that the computer is a high-speed digital processing engine that can perform a wide range of work, production or fun-related tasks. It's generally assembled from such building block components as a power supply, motherboard/central processing unit (CPU), random access memory (RAM), hard or solid-state disks and optional optical drives.

Most computers are designed to make use of peripheral hardware I/O or processing options that can be added to the system via hardware expansion slots that can be directly plugged into the system's motherboard or by connecting to the system via high-speed I/O protocols in order to expand the system to perform additional tasks that can't be handled by the computer's own hardware. Such serial I/O protocols (such as FireWire, USB or Thunderbolt) help to integrate the computer to the peripheral world of digital audio, MIDI, video, scanning, networking and tons of amazing things that haven't even been invented yet.

Software, that all-important binary computer-to-human interface, gives us countless options for performing all sorts of tasks such as digital audio production, MIDI sequencing, signal processing, graphics, word processing and music printing. Thanks to the countless software and shareware options that are available for handling and processing most types of media data, we can choose the platform OS, look, feel and programming style that best suits our own personal working style and habits. When you get right down to it, the fact that a computer can be individually configured to best suit the individual's own needs has been one of the driving forces behind the modern digital age. This revolution has turned the computer into a powerful "digital chameleon" that can change its form and function to fit the desired task at hand.

For the most part, two computer types are used in modern-day music production: the PC and the Mac. In truth, each brings its own particular set of advantages and disadvantages to personal computing, although their differences have greatly dwindled over the years. My personal take on the matter (a subject that's not even important enough to debate) is that it's a dual-platform world. The choice is yours and yours alone to make. Having said this, here are a few guidelines that can help you to choose which platform will best suit your needs:

- Which platform are you currently most familiar and comfortable with?
- Which platform and software is most used (and possibly demanded) by your clients?
- Which platform is most used by your friends or associates?
- What software and hardware do you already have and are familiar with?
- Which one do you like or feel best suits your needs?

Bottom line, do your research well and realize that the choices are totally up to you.

Like I said, it's a dual-platform world. Many professional software and hardware systems can work on either platform. As I write this, my music collaborator is fully Mac and I'm fully PC, and it doesn't affect our production process at all. Using both isn't much of a problem, either. Living a dual-platform existence can give you the edge of being familiar with both systems, which can be downright handy in a sticky production pinch.

In recent years, with the increased power of the laptop for either the Mac or PC platform, the question as to which type of computer system has become much less of a "how much power" question, than a personal choice question. Literally, in this day and age, the choice is up to you … as always with buying a system that will be your trusty side-kick for years to come, it's wise to make these system choices very carefully.

THE MAC

One computer type that's widely accepted by music professionals is the Macintosh® family of computers from Apple® Computer, Inc. The Mac® (Figure 8.1) offers up a graphical user interface (GUI) that uses graphic icons and mouse-related commands in a friendly environment that lets you move, expand, tile and stack windowed applications on the system's monitor screen. Due to the rigid hardware design constraints that are required by the Mac's operating system (OS), a tight integration between the system's hardware and software often exists that give many an "it just works" feeling.

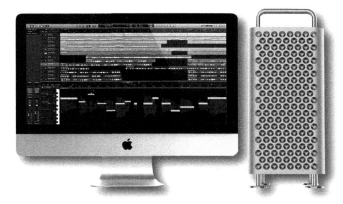

THE PC

Due to its cost effectiveness and sheer number of adopted systems in the home and business community, the Microsoft® Windows®-based personal computer (PC) clearly dominates the overall marketplace (musical or otherwise). Unlike the Mac®, which is made by one manufacturer, the PC's (Figure 8.2) general specifications were licensed out to the industry at large. Because of this, countless manufactures make compatible systems that can be factory- or user-assembled and upgraded using standard, off-the-shelf hardware components. Like the Mac® OS, Windows® is a sophisticated, graphic-based, multitasking environment that can run multiple task-based applications at a time. With the introduction of Microsoft's Win 10 OS, the differences between computer types with regard to hardware, software, networking, specs and peripheral integration have completely blurred.

FIGURE 8.1
The MacBook Pro and Mac Pro computers. (Courtesy of Apple Inc., www.apple.com.)

FIGURE 8.2
The PC. (a) CS450v5 Creation Station desktop PC. (Courtesy of Sweetwater, www.sweetwater.com.) (b) The Lenovo Legion Y740 laptop. (Courtesy of Lenovo, www.lenovo.com.)

(a)

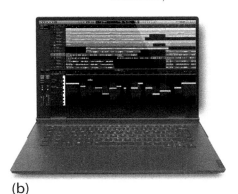

(b)

(a)

(b)

FIGURE 8.3
iOS-based DAW controllers.
(a) V-Control Pro DAW
controller. (Courtesy of
Neyrinck, www.vcontrolpro.
com.) (b) Cubase iC Pro
(Courtesy of Steinberg Media
Technologies GmbH, a division of Yamaha Corporation,
www.steinberg.net.)

THE iOS

Of course, in addition to the main two computing platforms, in music production (and many, many other production industries), the iOS must be added to
this list. Offering up cost-effective tools and apps that can improve productivity
and system control is the hallmark of this third platform (Figure 8.3).

With the addition of the iPad and iPhone, we can also add portability, wireless-
ness and on-the-go power to this list. Although the actual list is almost endless, two
of my favorite examples of how the iOSoffers up cost-effective power relating to:

1. A DAW remote controller (in this case the Mackie Universal Control) that
    originally cost around $1,200 can be easily emulated in app form for the
    cost of a couple of cups of coffee, or less.
2. A remote performance controller (Touchable 2) offers up direct, real-time
    performance and parameter control over Ableton Live for a cost that can't
    be touched by any hardware at basically any price.

Obviously, this is just the tip of the iceberg regarding the amount of applications
that are available for these Apple-based devices. Further info on the iOS and its
functionality can be found in Chapter 10.

THE DIGITAL AUDIO WORKSTATION

Throughout the history of music and audio production, we've grown used to
the idea that certain devices were only meant to perform a single task: a recorder
records and plays back, a limiter limits and a mixer mixes. Fortunately, the age
of the computer has totally broken down these traditional lines in a way that
has created a breed of digital chameleons that can change their functional colors
as needed to match the task at hand. Along these same lines, the digital audio
workstation isn't so much a device as a systems concept that can perform a wide

range of audio production tasks with relative ease and speed. Some of the characteristics that can (or should be) offered by a DAW include:

- Integration: one of the biggest features of a workstation is its ability to provide centralized control over the digital audio recording, editing, processing and signal routing functions within the production system. It should also provide for direct communications with production-related hardware and software systems, as well as transport and sync control to/from external media devices.
- Communication: a DAW should be able to communicate and distribute pertinent audio, MIDI and automation-related data throughout the connected network system. Digital timing (word clock) and synchronization (SMPTE time code and/or MTC) should also be supported.
- Speed and flexibility: these are probably a workstation's greatest assets. After you've become familiar with a particular system, most production tasks can be tackled in far less time than would be required using similar analog equipment. Many of the extensive signal processing, automation and systems communications features would be far more difficult to accomplish in the analog domain.
- Session recall: because all of the functions are in the digital domain, the ability to instantly save and recall a session and to instantly undo a performed action becomes a relatively simple matter.
- Automation: the ability to automate almost all audio, control and session functions allows for a great degree of control over almost all of a DAW's program and session parameters.
- Expandability: most DAWs are able to integrate new and important hardware and software components into the system with little or no difficulty.
- User-friendly operation: an important element of a digital audio workstation is its ability to communicate with its central interface unit: you! The operation of a workstation should be relatively intuitive and shouldn't obstruct the creative process by speaking too much "computerese".

I'm sure you've gathered from the above points that a digital audio workstation (and its associated hardware) which is capable of integrating audio, video and MIDI under a single, multifunctional umbrella can be a major investment, both in financial terms and in terms of the time that's spent learning to master the overall program environment. When choosing a system for yourself or your facility, be sure to take the above considerations into account. Each system has its own strengths, weaknesses and particular ways of working. When in doubt, it's always a good idea to research the system as much as possible before committing to it. Feel free to contact your local dealer for a salesroom test drive, or better yet, try the demo. As with a new car, purchasing a DAW (Figures 8.4–8.6) and its associated hardware can be an expensive proposition that you'll probably have to live with for a while. Once you've taken the time to make the right choice for you, you can get down to the business of making music.

FIGURE 8.4
Pro Tools hard-disk editing workstation for the Mac or PC. (Courtesy of Avid Technology, Inc., www.avid.com.)

SYSTEM INTERCONNECTIVITY

In the not-too-distant past, installing a device into a computer or connecting between computer systems could've easily been a major hassle. With the development of plug-n-play serial protocols, such as USB and Thunderbolt, hardware devices such as mice, keyboards, cameras, audio interfaces, MIDI interfaces, external drives, video monitors, portable fans, LED Christmas trees and cup warmers … can be plugged into an available port, installed and be up and running in no time … generally without a hassle. Additionally, with the development of a standardized network and Internet protocol, it's now possible to link computers together in a way that allows for the fast and easy sharing of data throughout a connected system. Using such a system, individuals and businesses alike can easily share and swap files with people on the other side of the world over the Web with relative ease.

FIGURE 8.5
Cubase Media Production System for the Mac or PC. (Courtesy of Steinberg Media Technologies GmbH, a division of Yamaha Corporation, www.steinberg.net.)

USB

In recent computer history, few interconnection protocols have affected our lives like the universal serial bus (USB). In short, USB is an open specification for

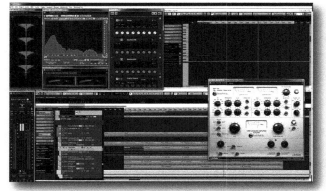

connecting external hardware devices to the personal computer, as well as a special set of protocols for automatically recognizing and configuring them. Here are the current USB specs:

FIGURE 8.6
Logic DAW for the Mac.
(Courtesy of Apple Inc.,
www.apple.com.)

- USB 2.0 (up to 480 megabits/sec = 60 megabytes/sec): for high throughput and fast transfer over the original USB 1.0 spec.
- USB 3.0 (up to 5 gigabits/sec = 640 megabytes/sec): for even higher throughput and fast transfer of the above applications.
- USB 3.1 (up to 10 gigabits/sec = 1.28 gigabytes/sec): for even higher throughput and fast transfer of the above applications.
- USB C is not a protocol, but is an actual connector spec that includes a plug that can be inserted in either orientation. Quite often (but not always) this connector can handle data rates up to the USB 3.1 data spec and when the Thunderbolt logo is present at the host port, is often capable of data speeds approaching that of Thunderbolt 3.

The basic overall characteristics of USB include:

- Up to 127 external devices can be added to a system without having to open up the computer. As a result, the industry has largely moved toward a "sealed case" or "locked-box" approach to computer hardware design.
- Newer operating systems will often automatically recognize and configure a basic USB device that's shipped with the latest device drivers.
- Devices are "hot pluggable", meaning that they can be added (or removed) while the computer is on and running.
- The assignment of system resources and bus bandwidth is transparent to the installer and end user.
- USB connections allow data to flow bi-directionally between the computer and the peripheral.
- USB cables can be up to 5 meters in length (up to 3 meters for low-speed devices) and include two twisted pairs of wires, one for carrying signal data and the other pair for carrying a DC voltage to a "bus-powered" device. Those that use less than 500 milliamps (1/2 amp) can get their power

directly from the USB cable's 5-V DC supply, while those having higher current demands will need to be externally powered. USB C, on the other hand, can supply up to 20 volts or 100 watts through the data cable.

- Standard USB 1 through 3 cables generally have different connectors at each end. For example, a cable between the PC and a device would have an "A" plug at the PC (root) connection and a "B" plug for the device's receptacle. USB-C cables, on-the-other-hand, "can" be the same connector at both ends, with each side having no particular plugin orientation (yeah)!

Cable distribution and "daisy-chaining" are done via a data "hub". These devices act as a traffic cop in that they cycle through the various USB inputs in a sequential fashion, routing the data into a single data output line.

Thunderbolt

Originally designed by Intel (in collaboration with Apple) and released in 2011, Thunderbolt (Figure 8.7) combines the DisplayPort and PCIe bus into a single, serial data interface. A single Thunderbolt port can support a daisy chain of up to six Thunderbolt devices (two of which can be DisplayPort display devices), that can run at such high speeds as:

- Thunderbolt 1 (up to 10 gigabits/sec = 1.28 gigabytes/sec) makes use of the same connector as the Apple Mini Display Port (MDP).
- Thunderbolt 2 (up to 20 gigabits/sec = 2.56 gigabytes/sec) makes use of the same connector as the Apple Mini Display Port (MDP).
- Thunderbolt 3 (up to 40 megabits/sec = 5.12 megabytes/sec) makes use of a USB-C connector that can be plugged into any USB-C port (on either the Mac or PC) that sports the Thunderbolt logo.

FIGURE 8.7
Intel's Thunderbolt3 protocol makes use of the USB-C type connector. (Courtesy of Apple Inc., www.apple.com.)

Obviously, the most amazing thing about this "hot pluggable" protocol is its speed. Another bonus is that it not only conforms to the USB spec (so that USB devices can be hot-plugged into it), but also conforms to the PCIe bus spec, allowing many hardware peripheral devices to be connected onto the Thunderbolt bus.

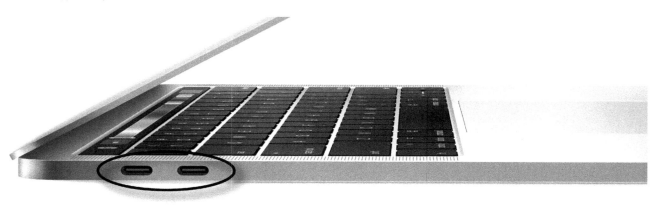

One important fact when entering into the world of Thunderbolt is the requirements, limitations and special needs that are often required when using these specialized cables. TB1 and 2 cables are often very specialized in design and spec, often with few things that can go wrong in practice. Using TB3, on the other hand, requires that you carefully choose your cabling in order to make a proper TB spec connection.

For example, when connecting two TB devices with only a standard USB-C data cable, you stand an excellent chance that no connection will be made. Even though the cable "looks" right, it won't have the necessary spec to make the connection. You will need to actually buy a cable that conforms to the TB3 spec. Additionally, passive cables (not containing data conditioning circuitry) that are greater than 0.5 meters will often not be able to pass data at the full 40 Mb/sec rate. Longer cables (up to 3 meters) must be high-quality active cables in order to get the job done. Passing the full TB3 40 Mb/sec rate at lengths greater than 3 meters can be done with the use of optical cables that conform to the TB3 spec. Basically, when setting up your system, it's wise to do your homework before buying just any ol' cable that might not work or severely limit your system's capabilities.

FireWire

Originally created in the mid-1990s as the IEEE-1394 standard, the FireWire protocol is similar to USB in that it uses twisted-pair wiring to communicate bidirectional, serial data within a hot-swappable, connected chain. Unlike USB (which can handle up to 127 devices per bus), up to 63 devices can be connected within a connected FireWire chain. FireWire most commonly supports two speed modes:

- FireWire 400 or IEEE-1394a (400 megabits/sec) is capable of delivering data over cables up to 4.5 meters in length. FireWire 400 is ideal for communicating large amounts of data to such devices as hard drives, video camcorders and audio interface devices.
- FireWire 800 or IEEE-1394b (800 megabits/sec) can communicate large amounts of data over cables up to 100 meters in length. When using fiber-optic cables, lengths in excess of 90 meters can be achieved in situations that require long-haul cabling (such as within sound stages and studios).

Unlike USB, compatibility between the two modes is mildly problematic, because FireWire 800 ports are configured differently from their earlier predecessor and therefore require adapter cables to ensure compatibility.

Audio over the Ethernet

One of the more recent advances in audio and systems connectivity in the studio and on stage revolves around the concept of communicating audio over the Ethernet (AoE). Currently, there are several competing protocols that range from being open-source (no licensing fees) to those that require a royalty to be designed into a hardware networking system.

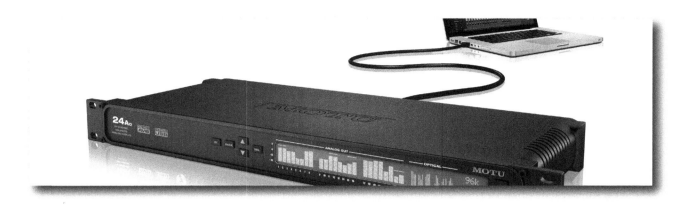

FIGURE 8.8
MOTU 24Ao audio interface with Audio Video Bridge (AVB) Switch and AVB Control app for communicating and controlling audio over Ethernet. (Courtesy of MOTU, Inc., www.motu.com.)

By connecting hardware devices directly together via a standard cat5 Ethernet cable (Figure 8.8), it's possible for audio channel counts of up to 512 × 512 to be communicated over a single connected network. This straightforward system is designed to replace bulky snake cables and fixed wiring within large studio installations, large-scale stage sound reinforcement, convention centers and other complex audio installations. For example, instead of having an expensive, multi-cable microphone snake run from a stage to the main mixing console, a single Ethernet cable could be run directly from a A/D mic/line cable box to the mixer (as well as the on-stage monitor mixer, for that matter) … all under digital control that often can include a redundant cable/system in case of unforeseen problems or failures.

In short, AoE allows for complex audio, MIDI and control system setups to be interconnected, digitally controlled and routed in an extremely flexible manner, and since the system is connected to the Internet, wireless control via apps and computer software.

THE AUDIO INTERFACE

An important device that deserves careful consideration when putting together a DAW-based production system is the digital audio interface. These devices can have a single, dedicated purpose, or they might be multifunctional in nature. In either case, their main purpose in the studio is to act as a connectivity bridge between the outside world of analog audio and the computer's inner world of digital audio (Figures 8.9–8.11). Audio interfaces come in all shapes, sizes and functionalities; for example, an audio interface can be:

- Built into a computer (although, more often than not, these devices are often limited in quality and functionality)
- A simple, two-I/O audio device
- A multichannel device, offering many I/Os and numerous I/O expansion options
- Fitted with one or more MIDI I/O ports

(a)

(b)

- One that offers digital I/O, word clock and various sync options
- Fitted with a controller surface (with or without motorized faders) that provides for direct DAW control integration
- Designed to include built-in DSP acceleration for offering additional plugin processing

These devices are most commonly designed as smaller stand-alone and/or 19″ rack mountable systems that plug into the system via USB, FireWire, Thunderbolt or AoE. An interface might have as few as two inputs and two outputs, or it might have more than 24. Recent units offer bit-depth/sample-rate options that range up to 24/96 or 24/192. In recent times, pretty much all interfaces will work with any DAW and platform (even Digidesign has dropped their use of proprietary hardware/software pairing).

Obviously, there are a wide range of options that should be taken into account when buying an interface. Near the top of this list (audio quality always being the top consideration) is the need for an adequate number of inputs and outputs (I/O). Although a number of interface designs include a large number of I/O channels, by far most have a limited I/O count that might offer access to additional I/O options should the need arise. This can include such options as:

- Lightpipe (ADAT) I/O, whereby each optical cable can give access to either eight channels at sample-rates of 44 or 48 k or four channels at 96 k (If

FIGURE 8.9

Portable audio interfaces: a. Steinberg UR22 MkII 2x2 audio interface (Courtesy of Steinberg Media Technologies GmbH, a division of Yamaha Corporation, www.steinberg.net.) b. Native Instruments Komplete Audio 6 audio interface. (Courtesy of Native Instruments GmbH, www.native-instruments.com.)

FIGURE 8.10

Presonus Studio 1824c 18x20 audio interface (front and back). (Courtesy of Presonus Audio Electronics, Inc., www.presonus.com.)

FIGURE 8.11
Apollo X Series audio interfaces with integrated UAD effects processing. (Courtesy of Universal Audio, www.uadio.com, © 2017 Universal Audio, Inc. All rights reserved. Used with permission.)

this option is available), when used with an outboard Lightpipe preamp (Figure 8.12).

- Connecting additional audio interfaces to a single computer. This is possible whenever several compatible interfaces can be detected by and controlled from a single interface driver.
- Using an audio over the Ethernet protocol and compatible interface systems, additional I/O can be added by connecting additional AoE devices onto the network and patching the audio through the system drivers.

It's always important to fully research your needs and possible hardware options *before* you buy an interface. Anticipating your future needs is usually never an easy task … but it can save you from future heartaches, headaches and additional spending.

FIGURE 8.12
Universal Audio 4-710d Four-Channel Mic Preamplifier. (Courtesy of Universal Audio, www.uaudio.com, © 2017 Universal Audio, Inc. All rights reserved. Used with permission.)

Audio Driver Protocols

Audio drivers are software protocols that allows data to be communicated between the system's software and hardware. A few of the more common protocols for communicating audio and related plugin data are:

- WDM: this driver allows compatible single-client, multichannel applications to record and play back through most audio interfaces using

Microsoft Windows. Software and hardware that conform to this basic standard can communicate audio to and from the computer's basic audio ports.

- ASIO: the Audio Stream Input/Output architecture (which was developed by Steinberg and offered free to the industry) forms the backbone of VST. It does this by supporting variable bit depths and sample rates, multichannel operation and synchronization. This commonly used protocol offers low latency, high performance, easy setup and stable audio recording within VST.

- MAS: the MOTU Audio System is a system extension for the Mac that uses an existing CPU to accomplish multitrack audio recording, mixer, bussing and real-time effects processing.

- CoreAudio: this driver allows compatible single-client, multichannel applications to record and play back through most audio interfaces using Mac OS X. It supports full-duplex recording and playback of 16-/24-bit audio at sample rates up to 96 kHz (depending on your hardware and CoreAudio client application).

In most circumstances, it won't be necessary for you to be familiar with these protocols – you'll just need to be sure that your software and hardware are compatible for use with the driver protocol that works best for you. Of course, further information can always be found on the respective companies' websites.

LATENCY

When discussing the audio interface as a production tool, it's important that we touch on the issue of latency. Quite literally, latency refers to the buildup of delays (measured in milliseconds) in audio signals as they pass through the audio circuitry of the audio interface, CPU, internal mixing structure and I/O routing chains. When monitoring a signal directly through a computer's signal path, latency can be experienced as short delays between the input and monitored signal. If the delays are excessive, they can be unsettling enough to throw a performer off time. For example, when recording a synth track, you might actually hear the delayed monitor sound shortly after hitting the keys (not a happy prospect) and latency on vocals can be quite unsettling. However, by switching to a supported ASIO or CoreAudio driver and by optimizing the interface/DAW buffer settings to their lowest operating size (without causing the audio to stutter or distort), these delay values can be reduced down to unnoticeable or barely noticeable ranges.

In response to the above problem, certain modern interface systems include a function called "direct monitoring", which allows the system to monitor inputs directly from the monitoring source in a way that bypasses the DAW's monitoring circuitry. The result is a monitor (cue) source that is free from latency, allowing the artist to hear themselves without the distraction of delays in their monitor path.

SOUND FILE FORMATS

A wide range of sound file formats exist within audio and multimedia production; however, the most common ones that are used in professional audio which don't use data reduction compression are:

- Wave (.wav): the Microsoft Windows format supports both mono and stereo files at a variety of bit and sample rates. WAV files contain PCM coded audio (uncompressed pulse-code modulation formatted data) that follows the Resource Information File Format (RIFF) spec, which allows extra user information to be embedded and saved within the file itself.
- Broadcast wave (.wav): in terms of audio content, broadcast wave files are the same as regular wave files; however, text strings for supplying additional information (most notably, time code data) can be embedded in the file according to a standardized data format.
- Apple AIFF (.aif or .snd): this standard sound file format from Apple supports mono or stereo, 8-, 16- and 24-bit audio at a wide range of sample rates. Like broadcast wave files, AIFF files can contain embedded text strings.
- FLAC (Free Lossless Audio Codec): a lossless, royalty-free codec that is used by producers and audiophiles alike. FLAC can handle a wide range of bit-depth and sample-rates, well up into the high-definition range, while being able to encode up to eight channels (for immersive audio encoding).

Sound File Sample and Bit Rates

While the sample rate of a recorded bitstream (samples per second) directly relates to the resolution at which a recorded sound will be digitally captured, the bit rate of a digitally recorded sound file directly relates to the number of quantization steps that are encoded into the bitstream. It's important that these rates be determined and properly set before starting a session. Further reading on sample and bit rate depths can be found in Chapter 12.

DAW CONTROLLERS

Originally, one of the more common complaints against most DAWs (particularly when relating to the use of on-screen mixers) is the lack of hardware control that gives the user direct, hands-on access. Over the years, this has been addressed by major manufacturers and third-party companies in the form of:

- Hardware DAW controllers
- MIDI instrument controller surfaces that can directly address DAW controls
- On-screen touch monitor surfaces
- iOS-based controller apps

It's important to note that there are a wide range of controllers from which to choose and just because others feel that the mouse is cumbersome doesn't mean that you have to feel that way; for example, I have several controllers in my own studio, but the mouse is still my favorite physical interface. As always, the choice of what works best is totally up to you.

Hardware Controllers

Hardware controller types (Figure 8.13) generally mimic the design of an audio mixer in that they offer slide or rotary gain faders, pan pots, solo/mute and channel select buttons, as well as full transport remote functions. A channel select button might be used to actively assign a specific channel to a section that contains a series of grouped pots and switches that relate to EQ, effects and dynamic functions … or the layout may be simple in form, providing only the most-often used direct control functions in a standard channel layout.

Such controllers range in the number of channel control strips that are offered at one time. They'll often (but not always) offer direct control over eight input strips at a time, allowing channels groups to be switched in groups of eight (1–8, 9–16, 17–24, etc.); any number of the grouped inputs can be accessed on the controller, as well as on the DAW's on-screen mixer. These devices will also often include software function keys that can be programmed to give quick and easy access to the DAW's more commonly used program keys.

Instrument Controllers

Since all controller commands are transmitted between the controller and audio editor via MIDI and device-specific MIDI SysEx messages (see Chapter 2), it only makes sense that a wide range of MIDI instrument controllers (mostly keyboard controllers) offer a wide range of controls, performance triggers and system functionality that can directly integrate with a DAW (Figure 8.14). The added ability of controlling a mix, as well as remote transport control, is a nice feature that places transport and basic mix controls nearby.

FIGURE 8.13
Hardware Controllers. (a) Presonus Faderport 16 DAW controller. (Courtesy of Presonus Audio Electronics, Inc., www.presonus.com.) (b) Behringer X-Touch Compact DAW controller. (Courtesy of Behringer., www.behringer. com.)

(a)

(b)

FIGURE 8.14
Komplete Kontrol S49 keyboard controller (Courtesy of Native Instruments GmbH, www.native-instruments. com.)

FIGURE 8.15
Touch screen controllers. (a) The Raven MTi Multi-touch Audio Production Console. (Courtesy of Slate Pro Audio, www.slateproaudio.com.) (b) V-Control Pro DAW controller for the iPad. (Courtesy of Neyrinck, www.vcontrolpro .com.)

Touch Controllers

In addition to the wide range of hardware controllers that are available on the market, an ever-growing number of software-based touch-screen monitor controllers (Figure 8.15a) have begun to take over the market. These can take the form of standard touch-screen monitors that let you have simple, yet direct control over any software commands, or they can include software that gives you additional commands and control over specific DAWs and/or recording-related software in an easy-to-use fashion. Since these displays are computer-based devices themselves, they can change their form, function and entire way of working with a single … uh, touch.

In addition to medium-to-large touch control screens, a number of Wi-Fi-based controllers are available for the iPad (Figure 8.15b). These controller "apps" offer direct control over many of the functions that were available on hardware controllers that used to cost hundreds or thousands of dollars … but are now emulated in software and can be purchased from an app "store" for the virtual cost of a cup of coffee.

(a)

(b)

AUDIO AND THE DAW

By their very nature, digital audio workstations are software programs that integrate with computer hardware and functional applications to create a powerful and flexible audio, MIDI and even visual production environment. These programs commonly offer extensive record, edit and mixdown facilities for such uses in audio production as:

- Extensive sound file recording, edit and region definition and placement
- MIDI sequencing and scoring
- Real-time, on-screen mixing
- Real-time effects
- Mixdown and effects automation
- Sound file import/export and mixdown export
- Support for video/picture playback and synchronization
- Systems synchronization
- Audio, MIDI and sync communications with other audio programs (e.g., ReWire)
- Audio, MIDI and sync communications with other effects and software instruments (e.g., VST technology)

This list is but a small smattering of the functional capabilities that can be offered by an audio production DAW.

Suffice it to say that these software production tools are extremely powerful and varied in their form and function. As you can see, even with their inherent strengths, quirks and complexities, their basic look, feel and operational capabilities have, to some degree, become unified among the major DAW competitors. Having said this, there are enough variations in features, layout and basic operation that individuals (from aspiring beginner to seasoned professional) will have their favorite DAW make and model. With the growth of the DAW and computer industries, people have begun to customize their computers with features, added power and peripherals that rival their love for souped-up cars and motorcycles. In the end, though (as with many things in life), it doesn't matter which type of DAW you use – it's how you use it that counts!

Sound Recording and Editing

Most digital audio workstations are capable of recording sound files in mono, stereo, surround or multichannel formats (either as individual files or as a single interleaved file). These production environments graphically display sound file information within a main graphic window (Figure 8.16), which contains drawn waveforms that graphically represent the amplitude of a sound file over time in a what you see is what you get (WYSIWYG) fashion. Depending on the system type, sound file length and the degree of zoom, the entire waveform may be

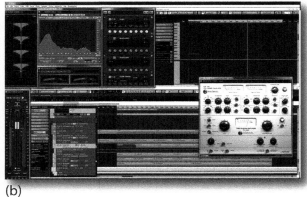

(a)　　　　　　　　　　　　　　　　　　　　　　　(b)

FIGURE 8.16
The main edit windows. (a) Avid's ProTools. (Courtesy of Avid Technology, Inc., www.avid.com.) (b) Steinberg Cubase audio production software. (Courtesy of Steinberg Media Technologies GmbH, a division of Yamaha Corporation, www.steinberg.net.)

shown on the screen, or only a portion will show … as it scrolls over the course of the song or program. Graphic editing differs greatly from the "razor blade" approach that's used to cut analog tape, in that the waveform gives us both visual and audible cues as to precisely where an edit point should be. Using this common display technique, any position, cut/copy/paste, gain or time changes will be instantly reflected in the waveforms on the screen. Almost always, these edits are nondestructive using a process whereby the original sound file isn't altered, only the way in which the region in/out points are accessed or the file is processed will be changed, undone, redone, copied, pasted, virtually without limit.

The nondestructive edit capabilities of a DAW refer to a disk-based system's ability to edit a sound file without altering the data that was originally recorded to disk. These important cut, copy, paste, edit and process capabilities mean that any number of edits, alterations or program versions can be performed and saved to disk without altering the original sound file data.

Real-Time, on-Screen Mixing

In addition to their ability to edit and define regions extensively, one of the most powerful cost- and time-effective features of a digital audio workstation is the ability to offer on-screen mixing capabilities (Figure 8.17), known as mixing "in the box". Essentially, most DAWs include a digital mixer interface that offers most (if not more) of the capabilities that are offered by larger analog and/or digital consoles – without the price tag and size. In addition to the basic input strip fader, pan, solo/mute and select controls, most DAW software mixers offer broad support for EQ, effects plugins (offering a tremendous amount of DSP flexibility), routing, spatial positioning (pan and often surround-sound positioning), total automation (both mixer and plugin automation), mixing and transport control from an external surface, support for exporting (bouncing) a mixdown to a file … the list goes on and on and on.

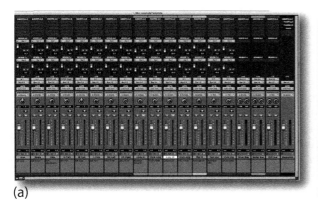
(a)

(b)

DSP EFFECTS

In addition to being able to cut, copy and paste regions within a sound file, it's also possible to alter a sound file, track or segment using digital signal processing techniques. In short, DSP works by directly altering the samples of a sound file or defined region according to a program algorithm (a set of programmed instructions) in order to achieve a desired result. These processing functions can be performed either in real time or non-real time (offline).

Virtually all workstations offer a number of stock DSP effects that come bundled with the program; however, a staggering range of third-party plugin effects can be found on-line or purchased that can be inserted into a signal path which perform functions for any number of tasks ranging from the straightforward to the wild-'n'-zany. These effects can be programmed to seamlessly integrate into a host DAW application that conforms to such plugin platforms as:

- Audio Units (AU): developed by Apple for audio and MIDI technologies in OS X; allows for a more advanced GUI and audio interface.
- Virtual Studio Technology (VST): a native plugin format created by Steinberg for use on either a PC or Mac; all functions of a VST effect processor or instrument are directly controllable and automatable from the host program.
- MOTU Audio System (MAS): a real-time native plugin format for the Mac that was created by Mark of the Unicorn as a proprietary plugin format for Digital Performer; MAS plugins are fully automatable and do not require external DSP in order to work with the host program.
- AudioSuite: a file-based plugin that destructively applies an effect to a defined segment or entire sound file, meaning that a new, affected version of the file is rewritten in order to conserve on the processor's DSP overhead (when applying AudioSuite, it's often wise to apply effects to a copy of the original file so as to allow for future changes).
- Real-Time Audio Suite (RTAS): a fully automatable plugin format that was designed for various flavors of Digidesign's Pro Tools and runs on the power of the host CPU (host-based processing) on either the Mac or PC.

FIGURE 8.17
DAW on-screen mixer. (a) Avid's ProTools. (Courtesy of Avid Technology, Inc., www. avid.com.) (b) Steinberg Cubase audio production software. (Courtesy of Steinberg Media Technologies GmbH, a division of Yamaha Corporation, www.steinberg.net.)

- Time Domain Multiplex (TDM): a plugin format that can only be used with Digidesign Pro Tools systems (Mac or PC) that are fitted with Digidesign Farm cards; this 24-bit, 256-channel path integrates mixing and real-time digital signal processing into the system with zero latency and under full automation.

These popular software applications (which are designed and programmed by major manufacturers and small, third-party startups alike) have helped to shape the face of the DAW by allowing us to pick and choose the plugins that best fit our personal production needs. As a result, new companies, ideas and task-oriented products are constantly popping up on the market, literally on a monthly basis.

In this day and age, the CPU of most computer systems will generally have sufficient power and speed to perform all of the DSP effects and processing needs of a project. Under extreme production conditions, however, the CPU might run out of computing steam and choke during real-time playback. Under these conditions, there are a couple of ways to reduce the workload on a CPU: on one hand, the tracks could be "frozen", meaning that the processing functions would be calculated in non-real time and then written to disk as a separate file. On the other hand, an accelerator card (Figure 8.18) that's capable of adding extra CPU power can be added to the system, giving the system extra processing power to perform the necessary effects calculations. Note that in order for the plugins to take advantage of the acceleration they need to be specially coded for that specific DSP card. Of course, as computers have gotten faster and more powerful, native processing packages have come onto the market, which make use of the modern computers multi-processor capabilities.

FIGURE 8.18
The UAD Satellite Thunderbolt 3 DSP accelerator for Mac and Win. (Courtesy of Universal Audio, www.uaudio.com, © 2020 Universal Audio, Inc. All rights reserved. Used with permission.)

The following section on processing effects describes only a few of the many possible DSP effects that can be plugged into the signal path of DAW.

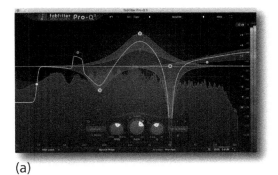

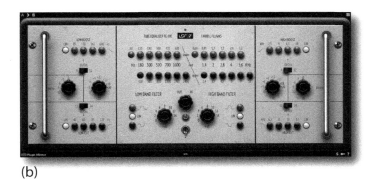

(a)   (b)

Equalization

EQ is, of course, a feature that's often implemented at the basic level of a virtual input strip (Figure 8.19). Most DAW "strips" also include one that gives full parametric control over the entire audible range, offering overlapping control over several bands with a variable degree of bandwidth control (Q). Beyond the basic EQ options, numerous third-party EQ plugins are available on the market that vary in complexity, musicality and market appeal.

Dynamics

Dynamic range processors (Figure 8.20) can be used to change the signal level of a program. Processing algorithms are available that emulate a compressor (a device that reduces gain by a ratio that's proportionate to the input signal), limiter (reduces gain at a fixed ratio above a certain input threshold) or expander (increases the overall dynamic range of a program). These gain changers can be inserted directly into a track, used as a grouped master effect or inserted into the final output path for use as a master gain processing block.

In addition to the basic complement of stock and third-party dynamic range processors, a wide assortment of multiband dynamic plugin processors (Figure 8.21) are available for general and mastering DSP applications. These processors allow the overall frequency range to be broken down into various frequency bands. For example, a plugin such as this could be inserted into a DAW's main output

FIGURE 8.19
EQ plugins. (a) Fabfilter-pro-q 3. (Courtesy of Fabfilter, www.fabfilter.com.) (b) Lindell Audio TE-100. (Courtesy of Lindell Plugins, www.lindellplugins.com.)

FIGURE 8.20
Compressor plugins. (a) API-2500 Compressor Plug-in. (b) Summit Audio TLA-100A Tube Leveling Amplifier. (Courtesy of Universal Audio, www.uaudio.com, © 2020 Universal Audio, Inc. All rights reserved. Used with permission.)

(a)   (b)

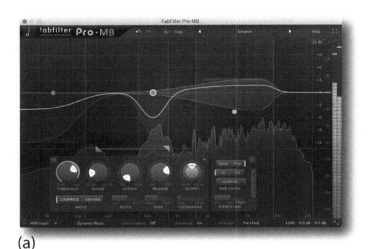

(a)

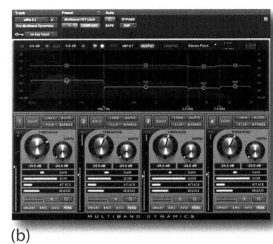

(b)

path, which allows the lows to be compressed while the mids are lightly limited
and the highs are de-essed to reduce harsh sibilance in the mix.

Delay

Another important effects category that can be used to alter and/or augment a
signal revolves around delays and regeneration of sound over time. These time-
based effects use delay (Figure 8.22) to add a perceived depth to a signal or
change the way that we perceive the dimensional space of a recorded sound.
A wide range of time-based plugin effects exist that are all based on the use of
delay (and/or regenerated delay) to achieve such results as:

- Delay
- Chorus
- Flanging
- Reverb

Pitch and Time Change

Pitch change functions make it possible to shift the relative pitch of a defined region or track either up or down by a specific percentage ratio or musical interval. Most systems can shift the pitch of a sound file or defined region by determining a ratio between the current and the desired pitch and then adding (lower pitch) or dropping (raise pitch) samples from the existing region or sound file. In addition to raising or lowering a sound file's relative pitch, most systems can combine variable sample rate and pitch shift techniques to alter the duration of a region or track. These pitch- and time-shift combinations make it possible for such changes as:

- Pitch shift only: a program's pitch can be changed while recalculating the file so that its length remains the same.
- Change duration only: a program's length can be changed while shifting the pitch so that it matches that of the original program.
- Change in both pitch and duration: a program's pitch can be changed while also having a corresponding change in length.

When combined with shifts in time (delay), changes in pitch make it possible for a world of time-based effects, alterations, tempo changes and more to be created. For example:

- Should a note be played that's out of (or the wrong) pitch … instead of going back and doing an overdub, it's a simple matter to zoom in on that note and change its pitch up or down, till it's right
- Should you be asked to produce a radio commercial that is 30 seconds long, and the producer says (after the fact) that it has to be 28 seconds … it probably is a simple matter to time stretch the entire commercial, so as to trim the 2 seconds off.
- The tempo of an entire song can be globally shifted in time or pitch, at the touch of a button, to change the entire key or tempo of a song.
- Should you import a musical groove that's of a different tempo than your session tempo, most DAWs will let you slip the groove in time, so that its tempo fits perfectly … simply by dragging the boundaries. Changing the grooves pitch is likewise a simple matter.
- Dedicated pug-ins can also be used to automatically tune a vocal or instrumental track (Figure 8.23), so that the intonation is corrected, smoothed out or exaggerated for effect.
- A process called "warping" can also be used to apply micro changes in musical timing (using time and pitch-shift processing) to fit, modify, shake up or otherwise mangle a section within a passage or groove. Definitely, fun stuff!

If you're beginning to get the idea that the wonderful world of pitch-shifting is an important production tool … you're right. However, there are definitely limits and guidelines that should be adhered to, or at least experimented with. For starters:

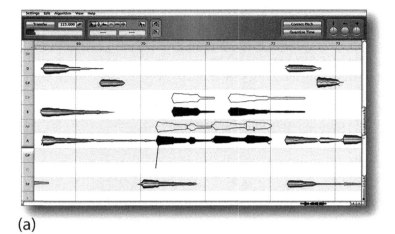

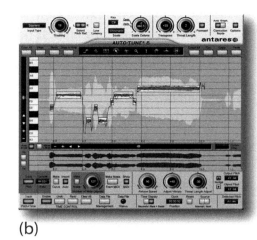

(a)   (b)

FIGURE 8.23
Automatic pitch correction software. (a) Melodyne Editor auto pitch correction system. (Courtesy of Celemony Software GmbH, www.celemony.com.) (b) Auto-Tune pitch correction system. (Courtesy of Antares Audio Technologies, www.antarestech.com.)

- A single program will often have several algorithms that can be applied to a passage (depending upon if it's percussive, melodic or continuous in nature). Not all algorithms are created equal. Also the algorithms of one program can easily sound totally different than those of another program. It isn't often straightforward or set-in-stone, as the processing is simply often too complex to predict … it generally requires experimentation and artistry
- Shifting in time or pitch (two sides of the same coin) by too great a value can cause audible side effects. You'll simply have to experiment in order to get the best results.

ReWire

ReWire and ReWire2 are special protocols that were co-developed by Propellerhead Software and Steinberg to allow audio to be streamed between two simultaneously running computer applications. Unlike a plugin, where a task-specific application is inserted "into" a compatible host program, ReWire allows the audio and timing elements of an independent client program to be seamlessly integrated into another host program. In essence, ReWire provides virtual patch chords that link the two programs together within the computer. A few of ReWire's supporting features include:

- Real-time streaming of up to 64 separate audio channels (256 with ReWire2) at full bandwidth from one program into its host program application
- Automatic sample accurate synchronization between the audio in the two programs
- An ability to allow the two programs to share a single soundcard or interface
- Linked transport controls that can be controlled from either program (provided it has some kind of transport functionality)

- An ability to allow numerous MIDI outs to be routed from the host program to the linked application (when using ReWire2)
- A reduction of the total number of system requirements that would be required if the programs were run independently

This useful protocol essentially allows a compatible program to be plugged into a host program in a tandem fashion. As an example, ReWire could allow Propellerhead's Reason (client) to be "ReWired" into Steinberg's Cubase DAW (host), allowing all MIDI functions to pass through Cubase into Reason while patching the audio outs of Reason into Cubase's virtual mixer inputs (Figure 8.24). For further information on this useful protocol, consult the supporting program manuals and Web videos.

Mixdown and Effects Automation

One of the great strengths of the "in the box" age is how easily all of the mix and effects parameters can be automated and recalled within a mix. The ability to change levels, pan and virtually control any parameter within a project makes it possible for a session to be written to disk, saved and recalled at a second's notice. In addition to grabbing a control and moving it manually (either virtually on-screen or from a physical controller), another interface style for controlling automation parameters (known as rubber-band or lane controls) lets you view, draw and edit variables as a graphic line that details the various automation moves over time.

As with any automation move, these rubber-band settings can be undone, redone or recalled back to a specific point in the edit stage. Often (but not always), the moves within a mix can't be "undone" and reverted back to any specific point in the mix. In any case, one of the best ways to save (and revert to) a particular mix version (or various alternate mix versions) is simply to save a specific mix under a unique, descriptive session file title (e.g., gamma_ultraviolet_radiomix01.ses)

FIGURE 8.24
ReWire allows a client program to be inserted into a host program (often a DAW) so the programs can run simultaneously in tandem.

ReWire "host"

ReWire "client"

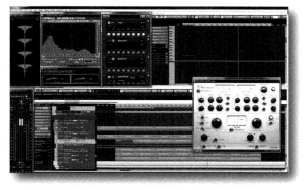

ReWire

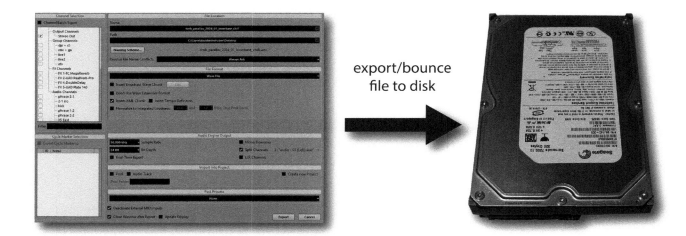

export/bounce
file to disk

FIGURE 8.25
Most DAWs can export
(bounce) session sound files,
effects and automation to a
final mixdown track.

and then keep on working. By the way, it's always wise to save your mixes on a regular basis (many a great mix has been lost in a crash because it wasn't saved or the auto-save function didn't work properly); in addition, progressively saving your mixes under various name or version numbers (mix01.ses, mix02.ses, etc.) can come in handy if you need to revert to a past version. In short, save often and save regularly!

Exporting a Final Mixdown to File

Once your mix is ready, most DAWs systems are able to export (bounce or print) part or all of a session mix to a single file or set of sound files (Figure 8.25). An entire session or defined region can be exported as a single interleaved file (containing multiple channels that are encoded into a single L-R-L-R sound file), or can be saved as separate, individual (L.wav and R.wav) sound files. Of course, a surround or multichannel mix can be likewise exported as a single interleaved file, or as separate files.

Often, the session can be exported in non-real time (a faster-than-real-time process that can include all mix, plugin effects, automation and virtual instrument calculations) or in real time. Usually, a session can be mixed down in a number of final sound file and bit/sample rate formats.

CHAPTER 9

MIDI and the DAW

Apart from electronic musical instruments, one of the most important music production tools that can be found in the modern-day project studio is the MIDI sequencer. Basically, a sequencer is a digital device or software application that's used to record, edit and output MIDI messages in a sequential fashion. These messages are generally arranged in a track-based format that follows the modern production concept of having instruments (and/or instrument voices) located on separate tracks. This visually interactive tool makes it easy for us humans to view MIDI data as tracks on a digital audio workstation (DAW) that follow along a straightforward linear time line.

These tracks (Figure 9.1) contain MIDI-related performance and control events that are made up of such channel and system messages as Note-On, Note-Off, velocity, modulation, aftertouch and program/ continuous controller messages. Once a performance has been recorded into sequenced tracks, they can be arranged, edited, processed, optionally exported as digital audio into the session and mixed into a musical performance. The overall MIDI data can then be saved as a file or within a DAW session and recalled at any time, allowing the data to be played back in itsoriginally recorded and edited order.

As is true with audio tracks on a DAW, sequenced tracks are designed to function in a way that's loosely similar to their distant cousin, the multitrack tape recorder. This gives us a familiar operating environment in which each instrument, set of layered instruments or controller data can be recorded onto separate, synchronously arranged tracks. Here, each track can be re-recorded, erased, copied and varied in level in a way that's far more flexible in its editing speed and control, offering all the cut-and-paste, signal-processing and channel-routing features that we've come to expect within a digital music production environment.

Here are just a few MIDI sequencing capabilities that are offered by a DAW:

- Increased graphics capabilities (giving us direct control over track and transport-related record, playback, mix and processing functions)
- Standard computer cut-and-paste edit capabilities

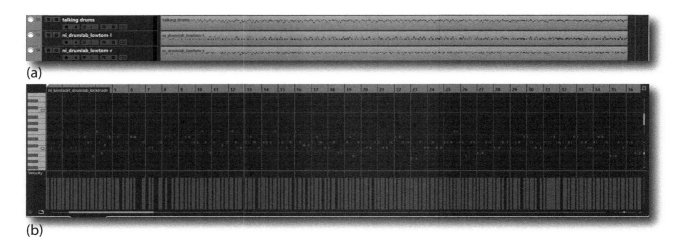

FIGURE 9.1
Examples of MIDI tracks
within a project. (a) MIDI
tracks. (b) Selected MIDI
track showing notes and
velocity levels.

- Ability to easily change note and controller values, one note at a time or over a defined range
- A window-based graphic environment (allowing easy manipulation of program and edit-related data)
- Easy adjustment of performance timing and tempo changes within a session
- Powerful MIDI routing to multiple ports within a connected system
- Graphic assignment of instrument voices via Program Change messages
- Ability to save and recall files using standard computer memory media

BASIC INTRODUCTION TO SEQUENCING

A MIDI sequence track is one that can be created within a DAW for the purpose of recording, editing and playing back MIDI performance and control-related data in a linear fashion. As with audio tracks, these systems use a working interface that's roughly designed to emulate a multi-track production environment.

Beyond using the traditional Record-Enable button to select the track or tracks that we want to record onto, all we need to do is:

- Select the MIDI input (source) port.
- Output (destination) port.
- MIDI channel (although most DAWS are also able to select all MIDI inputs as a source for ease of use).
- Instrument/plugin patch and other possible setup requirements.
- Finally, if you're routing to a hardware instrument (and even some software plugins), you'll want to create an audio track, route the instrument's audio to that track and then enable the audio monitoring function for that track.

Following these steps, you'll most likely be able to listen to the instrument as it's being played and then begin to lay down a track when you're ready … potentially, it can be that easy.

Setting a Session Tempo

When beginning a MIDI session, one of the first aspects to consider is the tempo and time signature (Figure 9.2). The beats-per-minute (bpm) value will set the general tempo speed for the overall session. This is important to set at the beginning of the session, so as to lock the overall "bars and beats" timing elements to this initial speed that's often essential in electronic and modern music production. This tempo/click element can then be used to lock the timing elements of other instruments and/or rhythm machines to the session (e.g., a drum machine plugin can be pulled into the session that'll automatically lock to the session's speed and timing).

- To avoid any number of unforeseen obstacles to a straightforward production, it's often wise to set your session tempo (or at least think about your options) "before" pressing the record button.
- Setting the initial tempo allows a click track to be used as a tempo guide throughout the session.
- This tempo/click element can then be used to lock the timing elements of other instruments and/or rhythm machines to the session (e.g., a drum machine plugin can be pulled into the session that'll automatically be locked to the session's speed and timing).
- When effects plugins are used, their delay and timing elements will be locked to a session.
- The session can be timed to visual elements (such as video and film), allowing sound cues and tempos to match the picture.

Failure to set a tempo might make it difficult or impossible for these advantages to be put to good use, as the tempo of a freewheeling performance will almost certainly not match or can drift from the defined tempo of a recorded MIDI session. This unfortunate event can severely limit the above options.

As with most things MIDI, when working strictly in this environment, it's extremely easy to change almost all aspects of your song (including the tempo) *provided* that you have initially worked to some form of an initial timing element (such as a base tempo). Although working without a set tempo can give a very human feel, it's easy to see how this might be problematic when trying to get

FIGURE 9.2
Example of tempo, time signature and metronome on/off selectors.

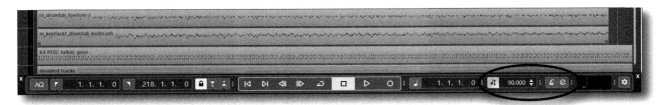

multiple MIDI tracks to work together with any form of musical sync and timing control. In truth, there are usually ways to pull corrective timing rabbits out of your technical hat … but the name of the game (as with most things recording) is forethought and preparation.

Changing Tempo

The tempo of a MIDI production can often be easily changed at any later time without worrying about changing the program's pitch or real-time control parameters. In short, once you know how to avoid potential conflicts and pitfalls, tempo variations can be made after the fact with relative ease. Basically, all you need to do is alter the tempo of a sequence (or part of a sequence) to best match the overall feel of the song. In addition, the tempo of a session can be dynamically changed over its duration by creating a tempo map that causes the bpm to vary by defined amounts at specific points over the course of a song. Care and preplanning should be exercised when a sequence is to be synced to another media form or device. To better understand the tempo map process, it's best to consult your DAW's manual.

Click Track

When musical timing is important (as is often the case in modern music and visual media production), a click track can be used as a tempo guide for keeping the performance as accurately on the beat as possible. A click track can be set to make a distinctive sound on the measure boundary or (for a more accurate timing guide) on the first beat boundary and on subsequent meter divisions (e.g., tock, tick, tick, tick, tock, tick, tick, tick, …). Most sequencers can output a click track by either using a dedicated beep sound (often outputting from the device or main speakers) or by sending Note-On messages to a connected instrument in the MIDI chain. The latter lets you use any sound you want and often at definable velocity levels. For example, a kick could sound on the beat, while a snare sounds out the measure divisions.

The use of a click track is by no means an absolute rule. A strong reason for using a click track (at least initially) is that it serves as a rhythmic guide that can improve the timing accuracy of a performance. However, in certain instances, it can lead to a performance that sounds stiff. For compositions that loosely flow and are legato in nature, a click track can stifle the passage's overall feel and flow. As an alternative, you could turn the metronome down, have it sound only on the measure boundary or listen through one headphone. As with most creative decisions, the choice is up to you and your current circumstance.

Care should be taken when setting the proper time signature at the session's outset. Listening to a 4/4 click can be disconcerting when the song is being performed in 3/4 time.

SETTING UP A TRACK

In keeping with the concept that "MIDI is not audio", the first steps toward setting up the system for recording a MIDI track to a sequencer is to make sure that both the MIDI and audio connections are properly made. If the MIDI connections between the instruments and interface are to be made via standard MIDI cables, you'll probably want to make sure that the MIDI In and MIDI Out cables are properly routed to and from the device ports. No matter what your setup, the audio pathways will also need to be connected. This might mean:

- Routing the audio outputs to an external audio mixer
- Routing the audio outputs to the inputs on a virtual DAW mixer (via the audio interface)
- Routing instrument MIDI cables and then assigning their ports/channels to the track
- Routing the virtual outputs of an instrument plugin to the virtual inputs of the DAW

Once the connections are made, simply arm the track, set the track to monitor the incoming MIDI activity and check that the MIDI data is flowing from the source to the target instrument (often by playing the source MIDI controller, checking the MIDI activity lights and then listening for sound). If there's no sound, check your connections and general settings (including your channel volume [controller 7] and instrument Local On/Off settings). If at first you don't succeed, be patient, trace through your system and settings and try, try again.

Now is a good time to talk about your controller or instrument Local On/Off settings. An instrument that has its "local" turned on means that whenever its keyboard (for example) is played, it will sound … no matter what. Turning the instrument's local off will often be advantageous when using a DAW sequencer, as the instrument will only respond to data that is being sent to it from the sequenced tracks. Put simply, if your main controller is also an instrument, you might consider turning local off, so that the sounds will only be played whenever the DAW's MIDI tracks are played.

Here's a tutorial that you might try on your DAW, so as to get familiar with the process:

1. Pull out a favorite MIDI instrument or call up a favorite plugin instrument.
2. Route the instrument's MIDI and audio cables (or plugin routing) to your DAW.
3. Create a MIDI track that can be recorded to.
4. Set the session to a tempo that feels right for the song.
5. Assign the track's MIDI input to the port that's receiving the incoming MIDI data.
6. Assign the track's MIDI output to the port and proper MIDI channel that'll be receiving the outgoing MIDI data to the instrument.
7. Create an audio track (if needed by the DAW), route the instrument's audio outs to this track and enable the input monitoring function.
8. If a click track is desired, turn it on (more about this later).
9. Name the track (*always* a good idea, as this will make it easier to identify the MIDI instrument/patch in the future).
10. Place the track into the Record Ready mode.
11. Play the instrument or controller. Can you hear the instrument? Do the MIDI activity indicators light up on the sequencer, track and MIDI interface? If not,

check your cables and run through the checklist again. If so, press Record and start laying down your track.

12. Once you've finished, you can jump back to the beginning of the recorded passage and listen to it.

From this point, you could then repeat the process and go about the process of laying down additional tracks (possibly with a different instrument or sound patch) until a song begins to form.

Multi-Track MIDI Recording

Although it's common for only one MIDI track to be recorded at a time, it's also totally possible to record multiple tracks at one time. This feature makes it possible for several instruments and/or performers to be recorded to several sequenced tracks in one live pass. For example, such an arrangement would allow for each trigger pad of a MIDI drum set to be recorded to its own track. Alternatively, several instruments of an on-stage electronic band could be captured to a sequence within a DAW during a live recording. The possibilities are completely up to you.

Punching in and out

MIDI tracks are capable of punching in and out of record while playing a sequence (Figure 9.3). This commonly used function lets you drop in and out of record on a selected track (or series of tracks) in real time in a way that mimics the traditional multitrack overdubbing process. Although punch-in and punch-out points can often be manually performed on the fly from the transport or often from a convenient foot pedal, most sequencers can also automatically perform a punch by graphically or numerically entering in the measure/beat points that mark the in and out location points. Once done, the sequence can be rolled back to a point a few measures before the punch-in point and the artist can play along while the sequencer automatically performs the necessary switching functions.

PUNCHING DURING A TAKE

1. Create a MIDI track in your DAW and record a musical passage. Save/name the session.

2. Roll back to the beginning of the take and play along. Manually punch in and out during a few bars (simply by pressing the REC button). Now, undo or revert back to your originally saved session.

3. Read your DAW manual and learn how to perform an automated punch (placing your

   punch-in and punch-out points at a logical place on the track).

4. Roll back to a point a few bars before the punch and go into the record mode. Did the track automatically place itself into record? Was that easier than doing it manually?

5. Feel free to try other features, such as record looping or stacking.

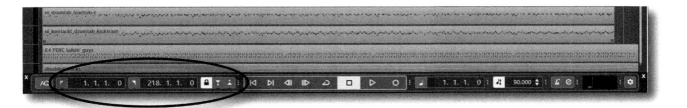

Pre-Count

A pre-count feature will sound a metronome measure (or defined count) before placing itself into the playback or record mode, giving the performer time to prepare for laying down his or her performed track. It's important that the proper tempo and time signature be entered into the session at the outset in order to get the right feel for the pre-count. Of course, a punch-in and punch-out will often be performed after the pre-count, making the process an automated one.

Record Loop

As the name suggests, the loop function in a DAW or sequencer lets you cycle-cycle-cycle between a predetermined in- and out-point in a continuous fashion. By placing the transport into a loop and then entering into record, it's possible to continuously cycle through a defined section in the Record mode, allowing the section to be repeated until the best take has been recorded (Figure 9.4). A pre-count can be placed at the loop's beginning, so as to give a preparatory break between takes.

Stacking

Certain DAW programs allow the above record loop function to be set up in such a way that each subsequent pass will be recorded to separate tracks in the session (Figure 9.5). This multiple take feature can be extremely useful in a number of ways, for example:

FIGURE 9.3
Example of punch-in/-out controls.

FIGURE 9.4
Example of a loop that is set to cycle between Measures 49 and 89.

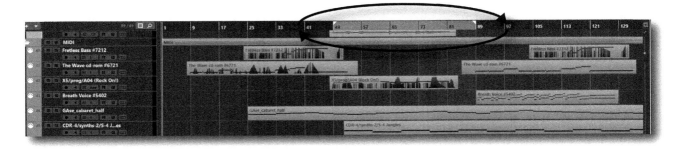

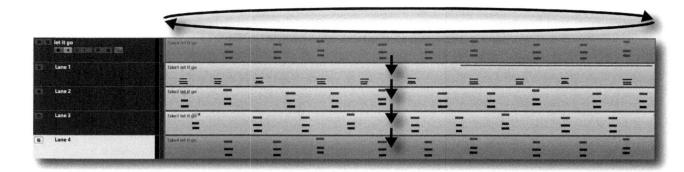

FIGURE 9.5
Stacking or cycling mode allows us to record each subsequent loop to a new track (or lane), so as to choose the best track (or combine parts into a composite track).

- When in the midst of laying down overdubs in the record-loop mode, it's often difficult to be objective about which take was the best. Stacking lets you decide at a later time.
- Multiple stacks can be combined to create a rich, layered effect on vocals, guitar – you name it.
- The best segments from various passes can be edited into a single, composite, winning take.

MIDI EDITING

One of the more important features that a sequencer (or MIDI track within a DAW) has to offer is its ability to edit sequenced tracks or blocks of performance/control data within a track. Of course, these editing functions and capabilities might vary from one DAW/sequencer to another. The main track window of a sequencer or MIDI track on a DAW is used to display such track information as the existence of track data, track names, MIDI port assignments for each track, program change assignments, volume controller values and other transport commands.

Depending on the sequencer, the existence of MIDI data on a particular track at a particular measure point (or over a range of measures) is indicated by the highlighting of track data in a way that's extremely visible. For example, in Figure 9.6, you'll notice that the MIDI tracks contain graphical bar display information. This means that these measures contain MIDI messages, while the non-highlighted areas don't.

FIGURE 9.6
Example of a MIDI track with recorded notes.

By navigating around the various data display and parameter boxes, it's possible to use cut-and-paste and/or edit techniques to vary note values and parameters for almost every facet of a musical composition. For example, let's say that we really screwed up a few notes when laying down an otherwise killer bass riff. With MIDI, fixing the problem is totally a no-brainer. Simply highlight each fudged note and drag it (them) to the proper note location – we can even change the beginning and endpoints in the process. In addition, tons of other parameters can be changed, including velocity, modulation and pitch bend, note and song transposition, quantization and humanizing (factors that can eliminate or introduce human timing errors that are generally present in a live performance), in addition to full control over program and continuous controller messages … the list goes on and on.

MIDI Track Editing

The main screen of a DAW will generally include a transport control surface, track lanes (which containing audio and MIDI-related automation functions) and other production-related controls. The MIDI track and its related windows are used to display such information as track-related data, track names, MIDI port assignments, program changes, controller values, mix and other related commands. An elapsed time and/or measure/beat display bar is placed within the edit window, allowing the user to easily zoom to a specific point within the session and to have a reliable time and measure reference. Depending on the display type, the existence of MIDI data on a track is generally indicated by small, dark blocks that appear in the track in a piano roll fashion.

Port assignments are usually made from the track window, allowing port and MIDI channels to be assigned for each MIDI track. The number of ports that are available will depend on the number of hardware interface and software instrument/applications that are available to the system.

PIANO ROLL EDITOR

One of the easiest and most common ways to see and edit MIDI track note values is through the use of the sequencer's piano roll edit window (Figure 9.7). This intuitive window lets you graphically edit MIDI data at the individual note level by displaying values on a continuous piano roll grid that's broken down into time (measures) and pitch (notes on a keyboard). Depending on where on the note's bar you click, you can:

- Change its beginning start time by clicking on the bar's front-end and dragging to the proper point.
- Change it to a new note number by vertically dragging it up or down the musical scale. Often, a keyboard is graphically represented at the window's left-hand side as a guide.
- Change its duration by clicking near the end of the bar and dragging to the proper point.

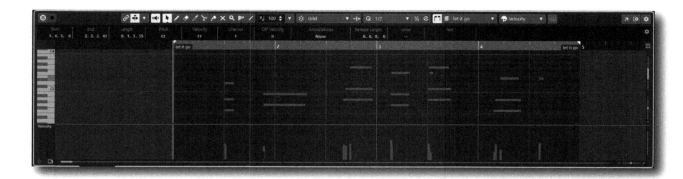

FIGURE 9.7
Example of a piano roll edit window.

As you might expect, the way in which notes can be edited might change from one sequencer to the next. The operation examples listed above are fairly common; however, you should consult your program's manual for further insights and details.

SNAP TO ...

Before continuing on, let's take a quick look at the concept of snapping an event to a defined boundary. In short, the snapping feature on a DAW or any type of sequencing system allows a MIDI, audio or other media type to magnetically jump to the closest defined time or measure boundary (Figure 9.8). For instance, it was said that we could click on the beginning of a note event within a piano roll display and move its beginning point in time. By turning on the "snap to measure" control, the beginning note would magnetically jump to the precise start point of the nearest measure. Of course, these snap boundaries can be defined in measures, measure subdivisions or time-related events. Again, now would be a good time to get out your trusty manual and read up on the snap feature.

SNAPPING TO

1. Record, open or import a MIDI track into your DAW.

2. Open the edit wind for that track (consult the manual; however, this is often done by double-clicking on the MIDI data track).

3. Zoom into the sequence, so the event bars can be easily dragged in time.

4. Turn off any snap functions.

5. Click in the beginning point of an event and drag it in time.

6. Turn on the "snap to measure" feature.

7. Click in the beginning point of an event and drag it in time. Does it snap to the nearest measure boundary?

8. Change the various snap settings and experiment with this useful feature.

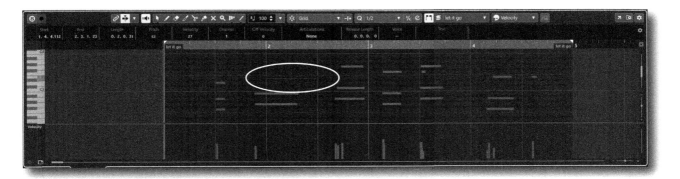

NOTATION EDITOR

Many sequencing packages allow notes to be entered and edited into a sequence using standard musical notation (Figure 9.9). This window works in much the same way as the piano roll editor, except that pitch, duration and other values are displayed and edited directly into measures as musical notes on any clef. One really nice by-product of this feature is that most programs also let us print a track's musical part in standard notation. This can be really useful for creating basic lead sheets (lyrics can even be entered into some sequences) for a particular instrument or part. Certain software packages are also able to print all of the tracks and parts of a sequence in a score form (where all the parts are arranged and printed onto the pages that can be reviewed and/or handed out to musicians during a session). Because the layout and printing demands are greater when laying out a printed score, many manufacturers offer a separate program for carrying out this specific production task.

Drum Pattern Entry

In addition to real-time and step-time entry, most sequencers will allow for drum and percussion notes to be entered into a drum pattern grid (Figure 9.10). This graphical environment is intuitive to most drummers and allows for patterns to be quickly and easily programmed and then linked together as a string of patterns that can form the backbone of a song's beat.

FIGURE 9.8
Example where the highlighted note has snapped to the half-measure points.

FIGURE 9.9
Example of a score edit window.

CREATING A DRUM PATTERN

1. Read your DAW's manual and learn how to open and use the basic functions of your system's drum editor.

2. Open a new session, create a MIDI track and assign its MIDI outs to a drum machine or percussion section of an instrument/plugin.

3. Open the Drum Edit window for that track (let's restrict our practice to a 4-bar pattern).

4. Create a kick pattern by clicking on the first beat of every bar. Did it sound when you clicked on the grid? When it played back?

5. If so, click the snare part on the first and third beats.

6. Hopefully, you've created a simple pattern that can be trimmed and looped within your basic rhythmic session.

7. Now, copy that pattern and make a few variations to the beat

8. Continue practicing and have fun!

For those who want to dive into a whole new world of experimentation and sonic surprises, here's a possible guideline: "If the instrument supports MIDI, record the performance data to a DAW MIDI track during each take". For example, you could also:

- Record the MIDI out of a keyboard performance to a DAW MIDI track. If there's a problem in the performance you can simply change the note (just after it was played or later), without having to redo the performance – or, if you want to change the sound, simply pick another sound.
- Record the sequenced track from a triggered drum set or controller to a set of DAW MIDI tracks.
- Fill out the sound of a MIDI guitar riff by doubling the MIDI track with another patch or chord augmentation.

FIGURE 9.10
Example of a drum pattern window.

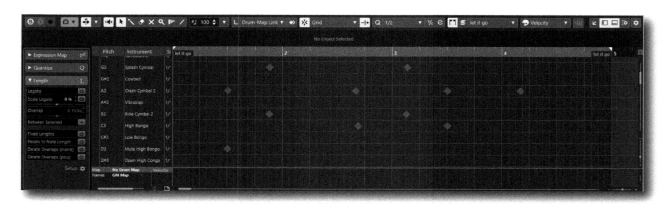

- Use a trigger device that can accept audio from a mic or recorded track to augment acoustic drum recordings by outputting the triggered MIDI messages to a sampler or instrument, in order to replace the bad tracks with samples that rock the house. By the way, don't forget to record the trigger outputs to a MIDI track or tracks, just in case you want to edit or change the sounds at a later time.
- Even if MIDI isn't involved, a drum replacement plugin could be used to replace bad sounds or to fill out a sound at a later time.

It's all up to you … as you might imagine, surprises can definitely come from experiments like these.

Event Editor

Although they're used less often than either the piano roll or notation edit windows, an event editor displays all of the MIDI messages that exist on a track (or in an entire sequence) as a sequential list of events. For example, Figure 9.11 shows a list of all MIDI events that occur in a MIDI track over time. In it, we can find information on when the note or event starts, stops, its volume and other editable values. Such an event window can be useful for instructing a sequencer to perform a task-based event (i.e., do something at such-and-such a time). For example, you could instruct the sequencer to trigger a specific sample at a specific time or insert a program change on a certain channel that tells an instrument to switch another instrument voice on cue.

Step Time Entry

In addition to laying down a performance track in real time, most sequenced tracks will let us enter note values into a sequence one note at a time. This feature (known as step time or step input) makes it possible for notes to be entered into a sequence without having to worry about their exact timing. Upon playback, the sequenced pattern will play back at the session's original tempo. The fact is, step entry can be an amazing tool, allowing a difficult or a blazingly fast passage to be meticulously entered into a pattern and then be played out or

FIGURE 9.11
Example of an event edit window.

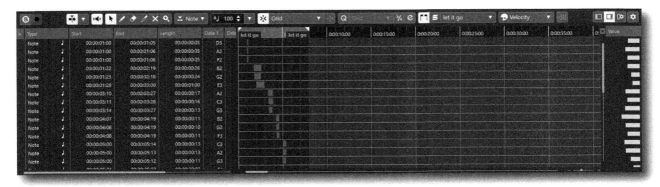

looped with a degree of technical accuracy that would otherwise be impossible for most of us to play. Quite often, this data entry style is used with fast, high-tech musical styles where real-time entry just isn't possible or accurate enough for the song.

PRACTICAL EDITING TECHNIQUES

When it comes to learning the Ins, Outs and Thrus of basic sequencing, absolutely nothing can take the place of diving in and experimenting with your own setup. Here, I'd like to quote Craig Anderton, who said: "Read the manual through once when you get the program (or device), then play with the software and get to know it 'before' you need it. Afterward, reread the manual to pick up the system's finer operating points". Wise words … although, personally, I tend to take another route and simply start pressing buttons and icons until I learn what I need. I honestly think there's something to be said for both approaches.

In the following sections, we'll be covering some of the basic techniques that'll help speed you on your way to sequencing your own music. Note that there are no rules to sequencing MIDI. As with all of music production (and the arts, for that matter), there are as many right ways to perform and play with making music via MIDI as there are musicians. Just remember that there are no hard and steadfast rules to music production – but there are always guidelines, tools and tips that can speed and improve the process.

Transposition

As was mentioned earlier, a sequenced MIDI track is capable of altering individual notes in a number of ways including pitch, start time, length and controller values. Changing the pitch of a note or the entire key of a song is extremely easy to do on a MIDI track. Depending on the system, a song can be transposed up or down in pitch at the global or defined measure level, thereby affecting the pitch or musical key of a song. Likewise, a segment can be shifted in pitch from the main edit, piano roll or notation edit windows by simply highlighting the bars and tracks that are to be changed and then dragging them or by calling up the transpose function from the MIDI edit menu.

Quantization

By far, most common timing errors begin with the performer. Fortunately, "to err is human", and standard performance timing errors often give a piece a live and natural feel. However, for those times when timing goes beyond the bounds of nature, an important sequencing feature known as quantization can help correct these timing errors. Quantization allows timing inaccuracies to be adjusted to the nearest desired musical time division (such as a quarter, 8th or 16th note). For example, when performing a passage where all involved notes must fall exactly on the quarter-note beat, it's often easy to make timing mistakes (even on a good day). Once the track has been recorded, the problematic passage can be highlighted and the sequencer can recalculate each note's start and stop times so they fall precisely on the boundary of the closest time division (Figure 9.12). Such quantization resolutions often range from full whole-note to 64th-note values and can also include triplet values.

Because quantization is used to "round off" the timing elements of a range of notes to the nearest selected beat resolution, it might be advisable to try to lock your playing in time with the sequencer's own timing resolution. This is done simply by selecting the tempo and time signature that you want and turning on the click track. By playing along with a click track, you're basically using the sequencer as a musical metronome. Once the track or song has been sequenced, the quantization function can be called up, which will further correct the selected note timing values.

In another attempt to extol the virtues of setting your initial tempo and click track settings, suppose you were to record the same sequence without a timing reference (most often a click track). In this case, the notes would be quantized to a timing benchmark that doesn't exist. That's not to say that it's impossible to quantize a sequence that wasn't played to a click – it's just trickier business. In short, whenever you feel you might need to quantize a segment, always give special consideration to a sequence's initial timing elements (i.e., using a click track).

QUANTIZATION

1. Read your DAW's manual and learn its quantization features.

2. Open a new session, create a MIDI track and assign its MIDI outs to an instrument or plugin.

3. Set the session to a tempo that feels right for the song and turn on the click track.

4. Record a small musical passage and save the file.

5. Zoom into a few of the notes and notice that they probably don't start exactly on the measure subdivisions.

6. Highlight the notes in the passage and set the quantize feature to a value that best matches the correct note length values (e.g., whole note, quarter note, 1/16th note).

7. Undo the change and try quantizing to various subdivision settings (e.g., whole note, half note, quarter note). Save the results to several session files.

Humanizing

When you get right down to it, one of the most magical aspects of music is its human ability to express emotion. A major factor used for conveying expression is the minute timing variations that are introduced during a performance. Whenever these variations become so large that they become sloppy, the first thing that many of us try is to tighten up the section's timing through quantization. The downfall of over-quantization, however, is that it can introduce

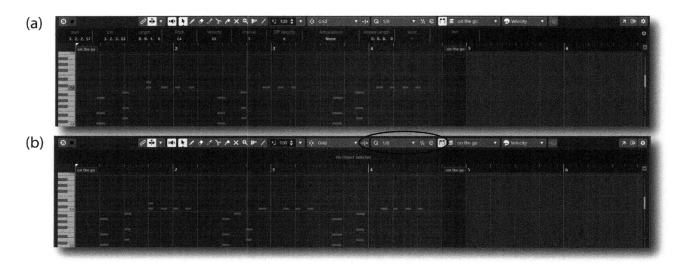

FIGURE 9.12

Example of quantization. (a) Non-quantized. (b) Quantized to the eighth-note internals.

a robotic accuracy that can take away from the basic human variations in the music, making it sound rigid and machine-like. One of the ways to reintroduce these variations in timing back into a quantized segment is through a process known as *humanization*.

The humanization process is used to randomly alter all of the notes in a selected segment according to such parameters as timing, velocity and note duration. The amount of randomization can often be limited to a user-specified value or percentage range and parameters and can be individually selected or fine-tuned for greater control. Beyond the obvious advantages, this process can help add expression by randomizing the velocity values of a track or selected tracks. For example, humanizing the velocity values of a percussion track that has a limited dynamic range can help bring it to life. The same type of life and human swing can be effective on almost any type of instrument. Let's give it a try.

HUMANIZING

1. Read your DAW's manual and learn its basic humanization features.

2. Open a quantized file from the above DIY on quantization.

3. Select a range of measures that have been quantized (preferably a series of fast, staccato notes).

4. Call up the humanize function and experiment with various time, velocity and length variables. (You can almost always undo your last move before trying a new set of values.)

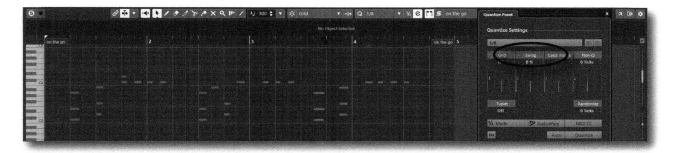

Swing

Another form of humanization, called *swing*, serves to introduce timing variables into the rhythmic feel of a passage by changing the timing of every other note in a musical part (or every second position in a pattern grid). This swing effect (Figure 9.13), which is generally expressed as a percentage value, gives a loose, shuffle-like feel to the pattern – often adding a degree of humanity to the track (if it's not overdone). You might want to swing out by trying the above DIY with the swing function.

Slipping in Time

Another timing variable that can be introduced into a sequence to help change the overall feel of a track is the *slip time* feature. Slip time is used to move a selected range of notes either forward or backward in time by a defined number of clock pulses. This has the obvious effect of changing the start times for these notes, relative to the other notes or timing elements in a sequence.

This function can be used to micro-tune the start times of a track so as to give them a distinctive feel. For example, nudging the notes of a sequenced percussion track forward in time by only a small number of clock pulses will effectively make the percussion track rush the beat, giving it a heightened sense of urgency or expectation. Likewise, retarding a track by any factor will give it a slower, backbeat kind of feel.

Slipping can also be used to move a segment manually into relative sync with other events in a sequence. For example, let's say that we've created a song that grooves along at 96 bpm; we've searched our personal archives and found a bridge (a musical break motif) that would work great in the piece. After inserting the required number of empty measures into the sequence and pasting the break into it, we found that the break comes in 96 clocks too late. No problem! We can simply highlight the break and slip it forward in time by 96 clocks. Often this process will take several tries and some manual nudging to find the timing that feels right, but the persistence could definitely pay off.

FIGURE 9.13
The swing function can offset each note by pre-determined and/or random amounts to make the music sound more human.

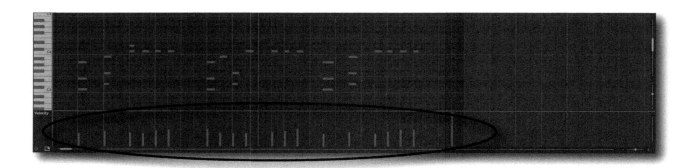

FIGURE 9.14
Example of editable controller values (in this case velocity).

FIGURE 9.15
Controllers can be used for direct control over software controller values. (a) Launch Control XL. (Courtesy of Novation Music, a brand of Focusrite Audio Engineering Plc, www.novationmusic.com.) (b) ATOM Production and performance pad controller. (Courtesy of Presonus Audio Electronics, Inc., www.presonus.com.)

Editing Controller Values

Almost every sequencer package allows controller message values to be edited or changed, and they often provide a simple, graphic window whereby a line or freeform curve can be drawn that graphically represents the effect that relevant controller messages will have on an instrument or voice (Figure 9.14). By using a mouse or other input device, it becomes a simple matter to draw a continuous stream of controller values that correspondingly change such variables as velocity, modulation, pan, etc.

With the advancement of software workstations, music production software and plugin applications, control over MIDI controller messages has become fully integrated into the basic mixing and sequencing applications, such that all you have to do is assign the control to a specific parameter and then twiddle the knob, move the fader or graphically draw the variables on-screen in a "what you see is what you get" (WYSIWYG) fashion (Figure 9.15).

(a)

(b)

Some of the more common controller values that can affect a sequence and/or MIDI mix values include the following (for the full listing of controller ID numbers, see Table 2.2 in Chapter 2):

- 1 Modulation Wheel, Lever or control   0–127
- 4 Foot Controller   0–127
- 7 Channel Volume (formerly Main Volume)   0–127
- 10 Pan   0–127
- 11 Expression Controller   0–127
- 64 Damper Pedal On/Off (Sustain)   <63 off, >64 on

It almost goes without saying that a range of controller events can be altered on one or more tracks by allowing a range of MIDI events to be highlighted and then altered by entering in a parameter or processing function from an edit dialog box. This ability to define a range of events often comes in handy for making changes in pitch/key, velocity, main volume, and modulation (to name a few).

CHANGING CONTROLLER VALUES

1. Read your DAW's manual and learn its basic controller editing features.

2. Open or create a MIDI track.

3. Select a range of notes or measures and change their Channel Volume settings (controller 7) over time. Does the output level change over the segment?

4. Highlight the segment and reduce the overall Channel Volume levels by 25%.

5. Take the segment's varying controller settings and set them to an absolute value of 95. Did that eliminate the level differences?

6. Now, refer to your DAW/sequencer manual for how to scale MIDI controller events over time.

7. Again, rescale the Channel Volume settings so they vary widely over the course of the segment.

8. Highlight the segment and scale the velocity values so they have a minimum value of 64 and a maximum of 96. Could you see and hear the changes?

9. Again, highlight the segment and instruct the software to fade it from its current value to an ending value of 0. Hopefully, you've just created a Channel Volume fade. Did you see the MIDI channel fader move?

10. Undo and start the fade with an initial value of 0 and a current value of 100% ending. Did the segment fade in?

Defining a range of note events can also be useful for scaling controller values (upward, downward or over time as a faded event change). For example, you could define a segment and place a minimum and maximum limit on the velocity values, effectively limiting the track's overall dynamic range (without using a compressor). You can also process a range of velocity or main volume messages so they ramp up or down over time, effectively creating a smooth fade-in or fade-out in the MIDI domain (not always a wise thing to do if you intend to transfer the MIDI track to an audio sound file, in which case it's generally best to make your changes in the audio domain).

Thinning Controller Values

Often, you'll find that these physically controlled or drawn curves will have their resolution set so high that literally hundreds of controller changes can be introduced into the sequence over just a few measures. In most cases, this resolution simply isn't necessary and might even create a data bottleneck when playing back a complex sequence. To filter out the gazillions of unnecessary messages that could be placed into a track, you might want to lower the control change resolution (if your sequencer and or controller hardware has such a feature) or you might want to thin the controller data down to a resolution that's more reasonable for the task at hand.

THINNING CONTROLLER VALUES

1. Read your DAW's manual and learn its basic controller thinning features.

2. Open or create a MIDI track.

3. Select a range of measures, choose an obvious controller (e.g., 7 [volume] or 1 [modulation]) and draw some complex and wacky curve.

4. Highlight the drawn curve, select the "thin controller data" function, and thin the data by about half. Did it remove tons of redundant controller messages?

5. Listen to the thinned track. Could you tell any difference in the sound or see the data visually in the controller edit window?

Filtering MIDI Data

Most sequencing applications are capable of filtering MIDI in a way that allows specific message or controller types within a data stream to be either recognized or ignored. This feature can come in handy in a studio or on-stage setting in that it can be used to block unwanted commands or controller types that might accidentally pass through. Depending on the program or hardware sequencer, a MIDI filter might be inserted into the entire sequencing system or on a track-by-track basis. In the latter instance, a filter could be inserted on one channel so as to block Program Change messages that might inadvertently be passed to an important on-stage instrument, while SysEx messages could be blocked on another data port/line so as to keep intense data bursts from clogging up a sensitive MIDI line.

Mapping MIDI Data

MIDI mapping is a process by which the value of a data byte or range of data bytes can be reassigned to another value or range of values. This function lets you change one or more of the parameters in an existing MIDI data stream to an alternate value or set of values. This process can be applied to a MIDI data stream to reassign channel numbers, transpose notes, change controller numbers and values, and limit controller values to a specific range. As with MIDI filtering, it's often possible to map data on a single channel so specific information can be mapped without affecting data on other channels. In this way, only a single device, chain of devices in a data line, or specific instrument voice will be affected.

Program Change Messages

Another type of automation that's supported by most sequencing packages involves the use of Program Change messages. As we saw in Chapter 2, the transmission of a Program Change message (ranging from 0 to 127) on a specific MIDI channel can be used to change the program or patch a preset number of an instrument or voice. By assigning a program change number to a specific sequence track, it becomes possible for an instrument (or a single part within a polyphonic instrument) to be automatically recalled to the desired patch setting. In short, an instrument or device patch can be automatically assigned to each track in a sequence, allowing the patch to be automatically recalled upon opening the file. Ideally, all you need to do is press the play button. Program changes can also be inserted in the middle of a sequence. Such a message can be used to instruct a synth voice to change from one patch to another in the middle of a song. For example, a synth could be used to play a B3 organ for the majority of a song; however, during a break, a Program Change message could instruct it to switch to an acoustic piano voice and then back again to finish the song. Patch changes such as these can also be used to change the settings for effects devices that can respond to MIDI program changes (i.e., changing a processor's settings from a rich room plate to a wacky echo setting in the middle of a sequence).

System Exclusive

As you may have guessed from comments in previous chapters, I'm a big fan of the power and versatility that System Exclusive can bring to almost every electronic musician's project studio. Most sequencers are able to read and transmit this instrument- and device-related data, allowing you to create your own sounds, grab sound patches from the Internet, swap patches with your friends or buy patch data disks from commercial vendors.

As we saw in Chapter 2, the System Exclusive message (or SysEx for short) makes it possible for MIDI manufacturers, programmers and designers to communicate customized MIDI messages between devices that talk MIDI. The general idea behind SysEx is that it uses MIDI to transmit and receive device-specific program, and patch data or real-time parameter information from one device to another. Basically, it's capable of turning an instrument or device into a digital chameleon. One moment an instrument can be configured with a certain set of sound patches and/or setup parameters and then, after having received a new SysEx data dump, you could easily end up with a whole new setup that's literally full of new and exciting (or not-so-exciting) sounds and settings. OK, let's take a look at a few examples of how SysEx can be put to good use.

Let's say that you have a Brand X Model Z synthesizer and it turns out that you have a buddy across town who also has a Brand X Model Z. That's cool, except your buddy's synth has a completely different set of sound patches ... and you want them! SysEx to the rescue! All you need to do is go over and transfer your buddy's patch data into your synth or into a MIDI sequencer as a SysEx data dump. To make life easier, make sure you take your instruction manual along (just in case you run into a snag) and follow these simple guidelines. I'll caution you that you're taking on these tasks at your own risk. Take your time – be patient and be careful during these procedures:

1. Before you do anything – *Back up your current patch data!* This can be done by transmitting a SysEx dump of your synthesizer's entire patch and setup data to your sequencer's SysEx dump utility or SysEx track on your sequencer (of course, you should get out both the device's manual and your sequencer's manual and follow their SysEx dump instructions *very carefully* during the process). This is so important that I'll say it again: back up your current patch data before attempting a SysEx dump! If you forget and download a new SysEx dump, your previous settings could easily be lost.

2. Save the data, according to your sequencer's manual.

3. Check that the dump was successful by reloading it back into the device in question. Did it reload properly? If so, your current patch data is now saved.

4. Next, connect your buddy's device to your sequencer. Dump these data to your sequencer. Save the new patch data (using a new and easily identifiable filename) according to your sequencer's manual and then safely back the data up.

5. Reconnect the sequencer to your synth and load the new data dump into it. Does your synth have a bunch of new sounds? Now reload your original SysEx dump back into your device. Are the original sounds restored?

If you were successful, you'll effectively have access to lots more sounds, the number of which will be limited solely by the number of SysEx dumps that you have hoarded for that particular device.

Another use for SysEx that many folks overlook is its ability to act as a backup medium for storing your patch and setup data just in case your system's memory gets corrupted or your instrument's RAM memory battery decides to go belly up. Here's an example of how a SysEx backup can save the day. A few years ago, I crashed the data in my WaveStation SR's memory by loading sounds from a WaveStation EX that wasn't totally compatible. Everything worked until I got to a certain patch and then the box went into a continual reboot loop – major freak-out time! No matter what I did, the system was totally locked up! The solution was to open up the box and take out the backup memory battery (which in turn cleared out the error as well as the box's internal RAM patch memory). All I had to do then was reload the factory SysEx patch settings back into the box and I was back in business in no time.

MIDI TRACKS

The creative process of manipulating, combining and playing with MIDI tracks to create an amazing musical experience is truly a technological artform that needs to be developed over time. A whole host of editing and creative tools are at your disposal to breathe more life into a sequenced track or set of tracks; a few of these are track splitting, track merging, voice layering and echoing.

Splitting Tracks

Although a MIDI track is encoded as a single entity that includes musical notes and other performance-related events, the data that's contained in a track can often be separated out in various creative ways and then split into separate tracks for transposition, further processing or routing to another instrument voice or multiple voices. As an example, let's say that we're working on a sequence that includes a synth part that was played live on stage. We, as co-producers, have decided that we'd like to have a greater degree of level mixing control over the right-hand part while leaving the bass part alone. This isn't really a problem. The best solution would be to split the track into two sequenced tracks by making an exact copy of the synth track and then pasting it onto a separate track that's assigned to a different MIDI channel and/or instrument. Once done, we can call up the piano roll window for the original track (allowing the note ranges to be easily seen), highlight the bass notes, and delete them. We can then repeat the process on the copied track and delete the upper notes (Figure 9.16). Now that the part has been split into two tracks, we can go about doing any number of things to them during postproduction. For example, we could:

- Change the channel volumes on each track to change their relative levels.
- Assign each part to a different instrument and/or voice.
- Transpose the upper notes or tag on a transposed harmony.
- Quantize the bass notes while leaving the upper notes untouched.

Merging Tracks

Just as a track can be split into two or more tracks, multiple tracks can be merged into a single track. On most sequencers, this can be done easily by highlighting the entire track or segment that you want to merge and copying it into memory. Once

FIGURE 9.16
Example of the original track being copied twice, and the notes are split between the high and low parts on the copied tracks. These can then be sent to different voices or layered for effect.

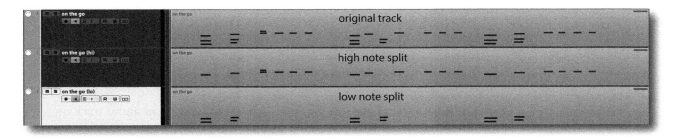

done, you can select the destination track and measure into which the copied data is to be merged and then invoke the sequencer's merge command. When merging MIDI data, you should keep in mind that it might be difficult to unmerge data that has been combined into a single track. For this reason, it's generally wise to keep the tracks separate and then simply assign them to the same MIDI channel and port or to save the original unmerged tracks as an archived backup.

Another use of track merging involves the playing back of an instrument track while recording real-time controller data from a pitch, modulation or other type of controller onto a separate track. This controller overdub process can help in keeping the original track intact, while real-time controller performances are recorded to a new track that's been assigned to the same port and channel. I often find that this process improves a performance, because it frees us up to concentrate on the performance. Once the controller tracks have been made to your liking, you can merge them with the original track (or simply keep them separate for future editing or archive purposes).

Layering Tracks

One of the more common and powerful tricks of the trade is the ability for instrument voices to be layered together to create a new, composite sound. Although digital samples and modern-day synthesis techniques have improved over the years, you're probably aware that many of these sounds will often have a character that can be easily distinguished from their acoustic counterparts. One of the best ways to fill out a sound and make it richer and more realistic is to layer multiple tracks together and then to assign each track to a different voice; when combined, these tracks will make a single, more complex and interesting voice. One example of layering would be to take the sound of a piano from synthesizer A and combine it with the sounds of a sampled Steinway from sampler B.

The process of layering can be carried out in several ways, for example:

- A layered track could be manually overdubbed onto another track and then mixed together as a combined voice.
- A single take could be duplicated to another track, assigned to another instrument voice and then mixed as a combined voice.
- A single take could be duplicated onto another track. This new track could then be assigned to another instrument or part and changes could be made (such as humanization and changes to note lengths). And, finally, combined sounds could be mixed together in a way that would create a richer, fuller sound.
- Layers can also be made by simply by assigning the output of a MIDI track to two or more instruments by transmitting the same data over multiple MIDI ports, channels and cables.

Of course, each layered part can be individually exported to an audio track or simply combined in the mix in ways that will achieve the best or most convincing results. For example, when working in surround sound, a full soundscape can be made by steering one stereo voice to the front speakers … and then steering a layered track to the rear.

MIDI Echo, Echo, Echo …

They say that you can never have too many effects boxes in your toy chest. Well, MIDI can also come to the rescue to help you set up effects in more ways than you might expect. For example, a part can be easily repeated in a digital delay fashion by copying a track to another track and then slipping that track (either forward or backward) in time. By assigning these tracks to the same destination (or by merging these tracks into one), you'll be setting up a fast and free MIDI echo effect. Simply repeat the process if you want to add more echoes.

MIDI Processing Effects

Just as signal processors can be used in an audio chain to create an effect by changing or augmenting an existing sound, a MIDI processor can often be inserted into a DAW MIDI track to alter its performance and processing functions, strictly in the MIDI domain. The following MIDI effects plugins are but a few that are included within Steinberg's Cubase/Nuendo DAW program (of course, you should check your DAW's own manual for MIDI plugin applications that might be included in its processing toolbox):

- Arpeggiator – a typical arpeggiator accepts a chord (organized group of simultaneous MIDI notes) as input and plays back each note in the chord separately in the playback order and speed set by the user.
- Autopan – this plugin works a bit like a low-frequency oscillator (LFO) in a synthesizer, allowing you to send out continuously changing and evolving MIDI controller messages (Figure 5.26). One typical use for this is automatic MIDI panning (hence, the name), but you can select any MIDI continuous controller event type.
- Chord Processing – a MIDI chord processor that allows complete chords to be assigned to single keys in a multitude of variations. There are three main modes of operation: Normal, Octave and Global.
- Compression – a MIDI compressor plugin is used for evening out or expanding differences in velocity. Though the result is similar to what you get with the velocity compression track parameter, the Compress plugin presents the controls in a manner more like regular audio compressors.
- MIDI Echo – a MIDI Echo that will generate additional echoing notes based on the MIDI notes it receives. It creates effects similar to a digital delay but also features MIDI pitch shifting and much more. As always it is important to remember that the effect doesn't echo the actual audio – it echoes the MIDI notes that will eventually produce the sound in the synthesizer.

AUDIO-TO-MIDI

Another tool that comes under the "You gotta try this!" category is the ability to take an audio file and extract the passage as relevant MIDI data. The power in creating an equivalent MIDI passage from a recorded groove or melody track is truly huge! For example, let's say that we have a really killer bass loop that has just the right vibe, but it lacks punch, is too noisy or just sounds wrong. A number of DAWs actually have algorithms that can detect and extract MIDI note and timing values that will result in a MIDI loop or sequence that can be saved as a separate MIDI file (Figure 9.17). As with all things MIDI, this new loop can be routed to a synth, sampler or anything to change the sound in a way that'll make it stand out. In short, you can do *anything* with the sequence … even call up a bass patch that has just that right punch for your needs. You owe it to yourself to check it out!

There is one thing to keep in mind when extracting MIDI from an audio track or loop, however. Each DAW will usually use different detection algorithms for drum/percussion, complex melody patterns and sustained pads. You might experiment to find which will work best with your track. Additionally, different DAWs will use different algorithms, which can easily yield completely different result. One DAW might work better than others on different source material. Experimentation and experience are often the name of the game here. Lastly, the results might be close to the original (in pitch and rhythmic pattern), but might require a bit of human editing. Once you get it right, the results can be truly powerful and fun to experiment with.

FIGURE 9.17
A number of DAWs are able to make use of algorithms to translate audio into a roughly analogous MIDI track. (Courtesy of Ableton AG, www.ableton.com.)

MIDI-TO-AUDIO

When mixing down a session that contains MIDI tracks, many folks prefer not to mix the sequenced tracks in the MIDI domain. Instead, they'll often export (bounce) the hardware or plugin instruments to an audio track within

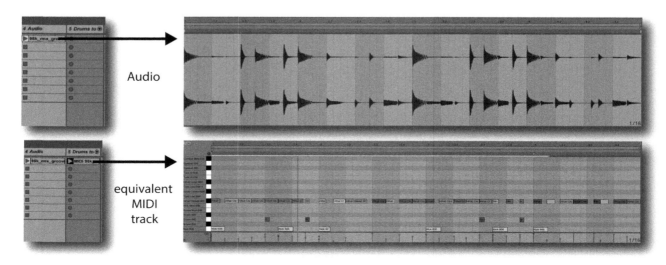

Audio

equivalent MIDI track

the project. Here are a few helpful hints that can make this process go more smoothly:

- Set the main volume and velocities to a "reasonable" output level, much as you would with any recorded audio track.
- Solo the MIDI and instrument's audio input track and take a listen, making sure to turn off any reverb or other effects that might be on that instrument track. If you really like the instrument effect, of course, go ahead and record it; however, you might consider recording the track both with and without effects, as you might want to make changes in mixdown.
- If any mix-related moves have been programmed into the sequence, you *might* want to strip out volume, pan and other controller messages before exporting the track. This can easily be done by making a copy of the existing track and then stripping the controller values from the copy … an export can be taken from this track.
- If a MIDI instrument has an acoustic element to it (such as a MIDI acoustic grand piano or MIDI guitar room/amp setup), you might consider recording the instrument to stereo or surround tracks in the studio and recording these tracks into the session. This will allow for greater flexibility should you need or want ambient "space" within the final mix.

It's worth mentioning again that it is always wise to save the original MIDI track or file within the session. This makes future changes in the composition infinitely easier. Failure to save your MIDI files will often limit your future options or result in major production headaches down the road.

Various DAWs will offer different options for capturing MIDI tracks as audio tracks within a session. As always, you should consult your DAW's manual for further details.

REPLACING AUDIO TRACKS VIA MIDI

In the world of recording production, it's widely known that one of the most difficult instruments to properly record is the drum set. "Thuddy" kicks, "ringy" snares, phase leakage – any and more of these problems can render a recorded drum track bad or even useless. Fortunately, MIDI can again come to the rescue by helping to resurrect a truly horrific drum track through the refined art of track replacement. Let's take a look at a few tools and tricks that can help save a project's butt:

- Acoustic triggers – for starters, acoustic or electro-acoustic pickups can be attached to a drum kit (that will also probably be acoustically miked for a session or on-stage event). These pickups can then be plugged into a drum module unit that's equipped with individual analog trigger inputs which, in turn, are able to take an input signal from each drum and instrument in the kit and convert this into MIDI messages that can be synchronously recorded to a MIDI track or tracks (either to MIDI tracks on a DAW or to a synchronized sequence track). Once the data have been encoded to MIDI, each drum sound can be altered, slipped in time, effected and used to replace the original sound.

- Triggering from track – just as the trigger inputs can be derived from a pickup, the source can also come from a previously recorded track. This means that a pitiful drum sound can be resurrected after the fact, allowing the offending track or tracks to be converted to MIDI, where the desired sounds can be triggered from a synth, sampler or plugin.
- Drum replacement software – in addition to the above methods, a more automated system exists in the form of a software plugin that can actually extract the originally recorded drum information and convert it into replacement tracks that can easily be placed into a DAW session (Figure 9.18). As is often the case with plugin technology, a wide range of replacement sounds and trigger timing options are available.

SAVING YOUR MIDI FILES

Just as it's crucial that we carefully and methodically back up our program and production media, it's important that we save our MIDI and session files while we're in production. This can be done in two ways:

- Periodically save your files over the course of a production.
- At important points throughout a production, you might choose to save your session files under new and unique names (mysong001, mysong002, etc.), thereby making it possible to easily revert back to a specific point in the production. This can be an important recovery tool should the session take a wrong turn or for re-creating a specific effect and/or mix. Personally, I save these session versions under a "mysong_bak" subdirectory within the project session.

One other precaution will often save your butt: always save your MIDI files within the session:

FIGURE 9.18
Steven Slate Drums Trigger 2 drum replacement plugin. (Courtesy of Steven Slate Drums, www.stevenslatedrums.com.)

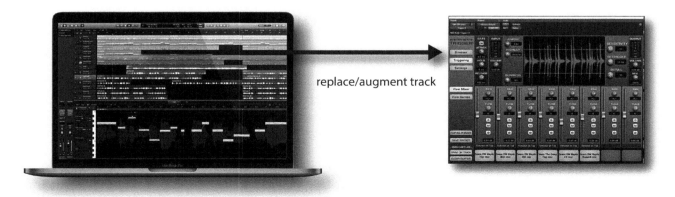

replace/augment track

- When working with MIDI within a DAW session, it's ALWAYS a good idea to keep and save the original MIDI tracks directly within the session. This makes it easy to go back and change a note, musical key or sounding voice or to make any other alterations you want. Not saving these files could lead to some major headaches or worse.
- Keeping your session MIDI files within a folder called "MIDI" will make it easier to identify and even hide (collapse) these tracks when they're not needed.

MIDI files can also be converted to and saved as a standard MIDI file for use in exporting to and importing from another MIDI program or for distributing MIDI data for use on the Web, to cell phones, etc. These files can be saved in either of two formats:

- Type 0: saves all of the MIDI data within a session as a single MIDI track. The original MIDI channel numbers will be retained. The imported data will simply exist on a single track. Note that if you save a multi-instrument session as a Type 0, you'll lose the ability to save the MIDI data to discrete tracks within the saved sequence.
- Type 1: saves all of the MIDI data within a session onto separate MIDI tracks that can be easily imported into a sequencer in a multi-track fashion.

DOCUMENTATION

When it comes to MIDI sequencing, one of the things to definitely keep in mind is the need for documenting any and all information that relates to your MIDI tracks. It's always a good idea to keep notes about:

- What instrument/plugin is being used on the track?
- What is the name of the patch?
- What are the settings that are being used (if these settings are not automatically saved within the session … and even if they are). In such cases, you might want to make a screenshot of the plugin settings, or if it's a hardware device, get out your cellphone or camera and take a picture that can be placed in the documentation directory.

Placing these and any other relevant piece of information into the session (within the track notes window or in a separate doc file) can really come in handy, when you have to revisit the file a year or so later and you've totally forgotten how to rebuild the track. Trust me on this one; you'll eventually be glad you did.

PLAYBACK

Once a sequence is composed and saved to disk, all of the sequence tracks can be transmitted through the various MIDI ports and channels to the instruments or devices to make music, create sound effects for film tracks, or control device parameters in real time. Because MIDI data exist as encoded real-time control commands and not as audio, you can listen to the sequence and make changes at any time. You could change the patch voices, alter the final mix, or change and experiment with such controllers as pitch bend or modulation – even change the tempo and key signature. In short, this medium is infinitely flexible in the number of versions that can be created, saved, folded, spindled and mutilated until you've arrived at the overall sound and feel you want. Once done, you'll have the option of using the data for live performance or mixing the tracks down to a final recorded media, either in the studio or at home.

During the summer, in a wonderful small-town tavern in the city where I live, there's a frequent performer who'll wail the night away with his voice, trusty guitar and a backup band that consists of several electronic synth modules and a laptop PC/sequencer that's just chock-full of country and western sequences. His set of songs for the night is loaded into a song playlist feature that was programmed into his sequencer. Using this playlist, he cues his sequences so that when he's finished one song, taken his bow, introduced the next song and compliments the lady in the red dress … all he needs to do is press the space bar and begin playing the next song. Such is the life of an on-the-road sequencer.

In the studio, using a DAW-based sequencer automatically integrates the world of audio and video with the MIDI production environment. Whenever more than one playback medium is involved in a production, a process known as synchronization is required to make sure that events in the MIDI, analog, digital and video media occur at the same point in time. Locking events together in time can be accomplished in various ways (depending on the equipment and media type used).

MIXING A SEQUENCE USING CONTINUOUS CONTROLLERS

Almost all DAW and sequencer types will let you mix a sequence in the MIDI domain using various controller message types. This is usually done by creating a physical or software interface that incorporates these controls into a virtual on-screen mixer environment. Instead of directly mixing the audio signals that make up a sequence, these controls are able to directly access such track controllers as Main Volume (controller 7), Pan (controller 10) and Balance (controller 8), most often in an environment that completely integrates into the workstation's overall mix controls. Since the mix-related data are simply MIDI controller messages, an entire mix can be easily stored within a sequence file. Therefore, it is completely possible to mix and remix your sequences with complete automation and total settings recall whenever a new sequence is opened. As is almost always the case with a DAW's audio and MIDI graphical user interface (GUI), the controller and mix interface will most always have moving faders and pots in a way that mimics most audio mixing control functions.

However, as you might be aware, there are several options that are available to you for dealing with both MIDI and audio automation within a project.

- As was just said, it's entirely possible to mix the MIDI tracks within a session completely in the MIDI domain using MIDI messages.
- It's also possible to not mix in the MIDI domain (using controller messages), but instead to mix the discrete audio channels that are being routed to the DAW completely in the audio domain.
- Lastly, it is often the case that the MIDI tracks will be exported to an audio track (with little or no MIDI mixing), whereby the audio can then be mixed directly using standard audio mixing practices. This later method means that the instruments can be recorded to an audio track, the original MIDI track should then be saved for future reference and finally the audio can be mixed using standard audio mixing practices, with little muss or fuss.

CHAPTER 10

The iRevolution

One of the most awe-inspiring developments in the modern age is the concept of mobile computing – specifically when it relates to the hundreds of millions of tablets, iPads and phones (cells, handys, mobiles or whatever your culture calls them). It has changed the way we communicate (or not communicate, if we don't occasionally look up to see what's around us) and the way that information is dealt with in an on-the-go, on-demand kinda way.

Of these mobile devices, the class of systems that has most affected the audio production community in a very direct way is the iOS range of devices from Apple. Those of us in recording, live sound and music production have been directly touched by these devices in ways that offer:

- Mobility: since iOS devices are wireless and portable by their very nature, these mini-computers allow us to record, generate and playback audio, perform complex control functions over a production computer, and use them as a mixer or production hardware device … our imagination is basically the only limit to what these devices can do from any untethered location.
- Affordability: quite often, these multi-function devices replace dedicated hardware systems that can cost up to a hundred times that of what an equally functional "app" might cost.
- Multi-functionality: given that these devices are essentially a computer, their capabilities are limited only by the functional programming that went into them. Most importantly, they have the ability to be a portable chameleon … changing form from being a controller, to a synthesizer, to a mixer, to a calculator, to a flashlight, to a phone and social networking device … all in a way that's always at your beck and call.
- Touch capabilities: the ability to directly interact with portable devices through the sense of touch (and voice) is something that we've all come to take for granted. This direct interaction allows all of the above advantages to be available at the touch of a virtual button.

Of course, it goes without saying that this chapter is simply a quick overview of a field that has grown into a major industry in and of itself. The number of apps, options and supporting connectivity and control hardware choices continues to grow on a monthly basis. To better find out what options are available and best suit your needs, I urge you to simply browse the App Store, search the Web for an ever-expanding number of online resources and wade your way through any number of YouTube videos and reviews on any iSubject that might be helpful to you and your studio scenario.

AUDIO AND THE iOS

Apple has had a long history of dealing with properly implementing audio into their operating systems (for the most-part). The major advantages of dealing with the iOS for portable computing and media players come down to two factors:

- The iOS has been developed to pass high-quality audio with a low amount of latency (system delay within the audio path).
- The operating system (including its audio implementation) is closed to outside third-party developers, allowing audio applications to be developed in a standardized way that properly makes use of Apple's programing architecture.

The following sections offer a basic glimpse into how the iOS can integrate into an audio production system with professional results that have made portable production more mobile, cost-effective and fun.

Core Audio on the iOS

Audio on the Mac OS and the iOS is handled through an integrated programming service called *Core Audio*. This audio architecture (Figure 10.1) is optimized for the computing resources that are available in a battery-powered mobile platform and is broken into application-level services which include:

- Audio Cue Services: used to record, playback, pause, loop and synchronize audio.

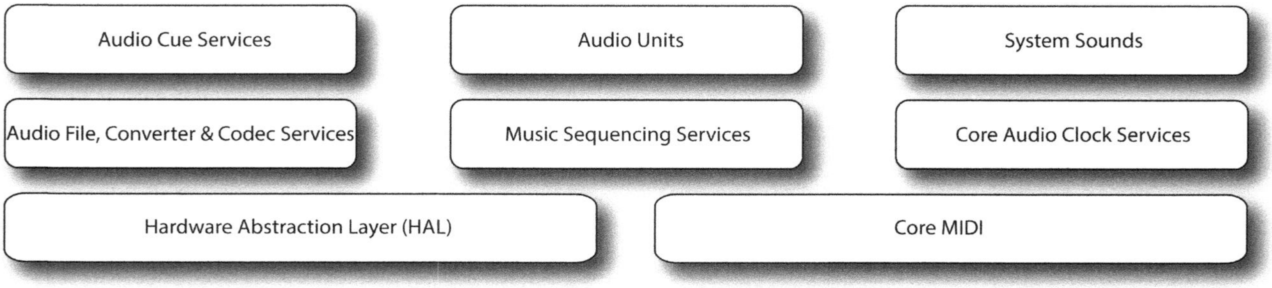

FIGURE 10.1
Core Audio's basic I/O architecture.

Audio Cue Services	Audio Units	System Sounds
Audio File, Converter & Codec Services	Music Sequencing Services	Core Audio Clock Services

Hardware Abstraction Layer (HAL)	Core MIDI

- Audio File, Converter and Codec Services: used to read and write data from disk (or media memory) and to perform audio file format transformations (in OS X, custom codecs can also be created).
- Audio Units: used to host audio units (audio plugins) that are available to an application.
- Music Sequencing Services: used to play MIDI-based control and music data.
- Core Audio Clock Services: used for audio and MIDI synchronization and time format management.
- System Sounds: used to play system sounds and user-interface sound effects.

AudioBus

AudioBus (Figure 10.2) is an iOS app that can be downloaded and used to act as a virtual patch cord for connecting together the ins, outs and thrus of various AudioBus compatible apps in a way that would otherwise not be possible. For example, an audio app that has been launched will obviously have its own input and an output. If we were to launch another audio app, there would be no way to "patch" the output of one app to the input of another within the iOS. Using AudioBus, it's now possible to connect a source device (input), route this through another effects device (which is not acting as "plugin", but as a stand-alone, processing app) ... and then patch the output of the effects app through to a final destination app (output).

Audio Units for the iOS

With the release of later versions of iOS, Apple has allowed for effects and virtual instrument plugins to be directly inserted into an app, much in the same way that plugins can be inserted into a DAW. This protocol, which is universally known across all Apple platforms as *Audio Units*, lets us download a plugin from

FIGURE 10.2
AudioBus showing input, effects and output routing. (a) Basic screen. (b) Showing the audio apps that are available to its input.

(a)

(b)

the App Store and then insert it directly into the processing path of an iOS DAW, video editor or other host app for DSP processing.

AUDIO CONNECTIVITY

Audio passes through an iOS device using Core Audio. Connecting to the device can be as simple as plugging into the mic/headphone out jack or Lightning port on the unit. Making a higher-quality, multi-channel audio connection will often require an external audio interface. The number of interface options for connecting an iPad or iPhone to a high-quality mic or integrating these devices into the studio system continues to grow in ways which generally fall into two categories:

- A docking device (Figure 10.3) can be used, that is specifically designed to connect an iOS device to the outside studio world of audio and MIDI.
- A standard audio interface that is class compliant (cc is a mode that allows the interface to be directly connected to an iOS device, using a readily available Apple Camera Adapter – as shown in Figure 10.4) can be connected to a device, allowing audio (and often MIDI) to pass in a normal I/O manner.

MIDI CONNECTIVITY

Using Core Audio, all iOS devices are capable of receiving and outputting MIDI without the need for additional software or drivers. All that's needed is a compatible I/O device that's capable of communicating with the device in either of two ways:

FIGURE 10.3
The Behringer iStudio dock for the iPad. (Courtesy of Behringer, www.behringer. com).

- By way of a docking device that can serve as an interface for audio and/or MIDI.
- By way of an Apple Camera Adapter that can connect the iOS device to a class-compliant MIDI or audio interface.

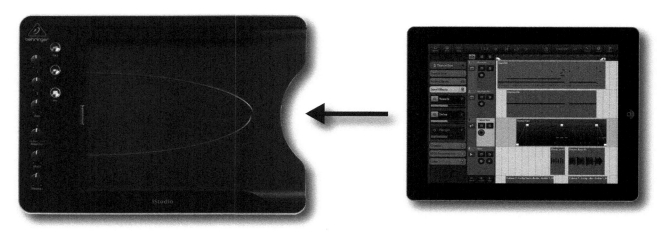

RECORDING USING iOS

Given the fact that iOS devices are capable of recording, editing, processing and outputting audio, it stands to reason that they would also excel at giving us access to production tools on a bus, in an airplane, by the pool or on our way to Mars … all in a cost-effective and versatile fashion.

Handheld Recording Using iOS

Another way that an iOS device can come in handy for recording audio is as a portable recording device. When used with a suitable mic accessory and recording app (Figure 10.5), an iPhone or iPad can be used to capture professional quality audio in an easy-to-use, on-the-go fashion, that can then be edited or transferred to a DAW for further editing, processing and integration into a project.

MIXING WITH iOS

Another way that the iOS has integrated itself into the recording and mixing process is through its pairing with newer hardware consoles or software DAWs (through the use of a wireless controller app as seen in Figure 10.6). This wireless control combination gives an engineer or producer unprecedented remote

FIGURE 10.4
A camera adapter can be used to connect a compliant device (such as an audio interface) to an iOS device.

FIGURE 10.5
Shure MV88 iOS handheld microphone. (Courtesy of Shure Incorporated, www .shure.com, Images © 2020, Shure Incorporated – used with permission.)

FIGURE 10.6
The Presonus StudioLive
16.0.2 Digital Recording
and Performance Mixer with
remote iPad app. (Courtesy of
Presonus Audio Electronics,
Inc., www.presonus.com.)

control over most mixing, panning, EQ, EFX and monitor control, either in the studio or on-stage.

iDAWs

Over the years, iPads have gotten powerful enough and apps have been written that allow an entire session to be transferred from our main workstation to a multitrack DAW on the iPad (Figure 10.7). It's easy to see (and hear) how such a powerful and portable device would be of benefit to the on-the-go musician or producer. He or she could record, mix and process a mix virtually anywhere. Offering up a surprising number of DSP options that are compatible with and can be read by a standard DAW system, anyone could put on a pair of in-ear monitors or noise-cancelling headphones and begin mixing … *virtually anywhere.*

FIGURE 10.7
Auria Pro DAW for the iPad.
(Courtesy of WaveMachine
Labs, Inc, www.wavema-
chinelabs.com.)

Taking Control of Your DAW Using the iOS

In addition to the many practical uses that are listed above, another huge contribution that iOS and mobile computing technology has made to music production is the iPad's ability to serve as a DAW remote control. For example, in the not-too-distant past, dedicated hardware DAW controllers would easily cost us over $1,000 and would need both a power and a wired USB connection. Now, with the introduction of iOS-based DAW controllers (Figure 10.8), the same basic functionality is available to us for less than $20 (if not for free). These apps allow the controller to fit in our hand and allow us to work wirelessly from any place in the studio, giving us the ability to:

- Mix from the studio, from your instrument stand, on-stage, in an audience … virtually anywhere.
- Remotely control all transport functions from anywhere within the facility.
- Control studio monitor and headphone sub-mixes from within the DAW (Figure 10.9), allowing the musician to control their own monitor mix, simply by downloading the app, connecting their iOS device to the studio network and then starting to create their own personal sub-mix.

Go ahead and search the App Store under "DAW controllers" and choose one that will best work for you (as always, it's important to do a bit of on-line research before buying, as that there can be differences and quirks that will set one app apart from the rest for your particular DAW). Once you've followed the install instructions to connect to your particular DAW, you'll be "remoting" around the studio from your padand/or phone in no time.

THE iOS ON STAGE

The iOS doesn't stop with the idea of controlling a DAW by using simple transport, control and mixing functions … when used with a performance-based DAW (such as Ableton Live), a number of iOS applications can be used to wirelessly integrate with a DAW allowing a performer to take control of their system to create a live performance set in real-time (Figure 10.10). Whereas, in the past, a hardware controller was used to perform with such a DAW in a way that had limited mobility and (more importantly) limited visual feedback, an

FIGURE 10.8
The DAW can be wirelessly controlled via an iOS DAW controller app. (Courtesy of Steinberg Media Technologies GmbH, a division of Yamaha Corporation, www.steinberg.net.)

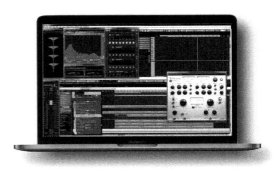

FIGURE 10.9

Musicians can easily mix their headphone sends remotely via an iOS DAW controller app. (Courtesy of Steinberg Media Technologies GmbH, a division of Yamaha Corporation, www.steinberg.net.)

iOS device can give the performer far better tactile and interactive response cues that allow the performer to clearly see exactly which audio and MIDI loops, effects and any number of performance controls are available to them … all in a way that can be interactively played, triggered and mutilated live, on-stage. As an electronic musician who works with such tools in an on-stage environment, I can tell you that iOS often changes the performer's experience into one that's far easier to see, understand and interact with over its hardware counterpart.

Another huge advancement in wireless control comes in the form of the iOS-based live sound controller systems (Figures 10.11). These devices are literally changing the way live sound hands are able to do their jobs. By giving a live sound mixer the freedom to mix wirelessly from anywhere in the venue, he or she can literally walk around the place and make any adjustments that are needed from the middle of the audience, front of house (FOH) position … again, virtually anywhere. Stage monitor mixing (which might use a completely separate mix or sub-mix from the main FOH mixer) can now also be accomplished wirelessly from a mix app. Depending upon the size and scale of the performance and venue, these stage mixes can be performed by a dedicated monitor mix person or by the performer themselves. As with all things wireless, it's all about freedom, mobility and flexibility.

FIGURE 10.10

Touchable 3 can be used to wirelessly control Ableton Live in a practice and performance setting. (Courtesy of Zerodebug, www.touch-able. net.)

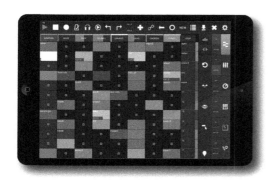

iOS AND THE DJ

The modern DJ is definitely no stranger to the power and portability of the iOS (Figure 10.12). Full sets can be easily pre-programmed or performed on-the-fly from an iPad or iPhone, especially when used in conjunction with external control hardware. Offering up most or all of the control of a laptop DJ system, these devices give the DJ full freedom to strut their stuff anywhere and anytime.

iOS AS A MUSICAL INSTRUMENT

Using iOS with the countless number of downloadable electronic instruments and music production tools allow us to quickly save and work on musical ideas anywhere, anytime and at a mere fraction of the cost of its hardware equivalent. They can also be used to integrate an instrument app into our working studio

FIGURE 10.11
StudioLive RML32AI wireless mixing system for live and recorded sound. (Courtesy of Presonus Audio Electronics, Inc., www.presonus.com.)

FIGURE 10.12
Traktor DJ2 DJ software for the laptop and iPad. (Courtesy of Native Instruments, Inc.; www.nativeinstruments.com.)

FIGURE 10.13
GarageBand for the Mac and iOS. (Courtesy of Apple Inc., www.apple.com.)

environment, adding rich and complex musical expression to a track (many of these apps aren't toys, but are the real deal).

Of course, this revolution started with GarageBand (Figure 10.13), a musical loop app that allows audio and MIDI to be dragged into a project timeline in a quick and easy way, without any musical experience. This app allowed musicians to first grasp the concept that the iOS could act as a serious musical instrument. It offers up a wide range of electronic loops and beats, as well as a set of virtual instruments (piano, guitar, bass, strings, etc.). These instruments can then be either sequenced from within the program itself, or (with the use of a MIDI interface connection) played from an external MIDI sequencer or controller source.

After GarageBand, individual developers and electronic instrument manufacturers quickly began to realize that there was a huge market for the recreation of classic synths, new synth designs, groove synths, beat generators and other electronic instrument types that range from being super-simple in their operation, to being sophisticated programs that equal or rival any hardware counterpart.

When using an iOS-compatible audio and MIDI interface (or simply a camera adapter/MIDI interface setup and the unit's 1/8″ headphone out jack), an iOS device can be paired with your DAW-based MIDI setup, literally allowing the instrument to be fully integrated into your musical system.

ABILITY TO ACCESSORIZE

Naturally, with the popularity of Apple and the iBrand, there are tons of accessories that can be added to any iDevice to customize its look and overall functionality. These can include:

- Desk stand adapters
- Protective covers
- Mics (intro to semi-pro in level)
- Docking stations
- Interface systems and adapters

Of course, this list could go on for quite some time. Suffice it to say, that if you wanted blue flames to shoot out of your device in a way that would spell out your name on the stage … just wait a few days and you'll be able to get it from the store for 99¢. It's all about customizing your system in a fun way that'll make the device uniquely yours.

CHAPTER 11

Sync

Over the years, audio production and electronic music have evolved into an indispensable production tool within almost all forms of media production. In video postproduction, for example, video workstations, digital audio workstations (DAWs), automated console systems and electronic musical instruments routinely work together to help create a soundtrack and refine it into its finished form. The technology that allows multiple audio and visual media to operate in tandem so as to maintain a direct time relationship is known as *synchronization* or *sync*.

SYNCHRONIZATION

Strictly speaking, synchronization occurs when two or more related events occur at precisely the same time. With respect to analog audio and video systems, sync is achieved by interlocking the transport speeds of two or more machines (Figure 11.1). For computer-related systems (such as digital audio, MIDI and digital video), synchronization between devices is often achieved through the use of a timing clock that can be fed through a separate line or can be directly embedded within the digital data signal itself. Frequently, it's necessary for analog and digital devices to be synchronized together; as a result, a number of ingenious forms of systems communication and data translation have been developed. In this chapter, the various forms of synchronization used between these analog and digital devices are discussed, as well as current methods for maintaining sync between the various media types.

Synchronization between Media Transports

Maintaining relative sync between media devices doesn't require that all transport speeds involved in the process be constant; however, it's more important that they maintain the same relative speed and position over the course of a program. Physical analog devices, for example, have a particularly difficult time achieving this. Due to differences in mechanical design, voltage fluctuations and tape slippage, it's a simple fact of life that analog tape-based devices aren't able to maintain a constant playback speed, even over relatively short durations. For this reason, accurate sync between analog and digital machines would be nearly

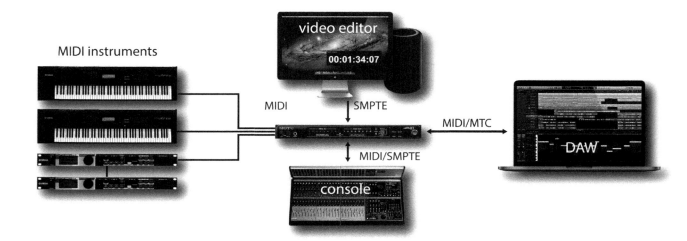

MIDI instruments

video editor

00:01:34:07

MIDI   SMPTE

MIDI/MTC

DAW

MIDI/SMPTE

console

FIGURE 11.1
Example of an integrated
audio production system.

impossible to achieve over any reasonable program length without some form
of timing lock.

Maintaining sync between digital devices (for the most part) is much easier to
achieve as there is no physical recording media involved that can cause variations
in speed. In addition, because the involved devices are digital (provided that
they can all communicate using a stable timing reference and communications
protocol) the process of maintaining sync "should" be relatively straightforward.

Over the course of this chapter, we'll discuss the various sync systems that have
been developed over the years to maintain a stable timing reference between
digital and analog audio and MIDI-related media devices.

TIME CODE

The standard method of interlocking audio, video and film transports makes use
of a code that was developed by the Society of Motion Picture and Television
Engineers (SMPTE, www.smpte.org). This time code (or SMPTE time code) iden-
tifies an exact position within recorded digital or analog media by assigning a
digital address that increments over the course of a program's duration. This
address code can't slip in time and always retains its original location, allowing
for the continuous monitoring of media position to an accuracy of between
1/24th and 1/30th of a second (depending on the media type and frame rates
being used). These divisional segments are called *frames*, a term taken from film
production. Each audio or video frame is tagged with a unique identifying number, known as a "time code address". This eight-digit address is displayed in the
form 00:00:00:00, whereby the successive pairs of digits represent hours:minute
s:seconds:frames or HH:MM:SS:FF (Figure 11.2).

The recorded time code address is then used to locate a position on hard disk,
magnetic tape or any other recorded media, in much the same way that a letter carrier uses a written address to match up, locate and deliver a letter to a

Time Display

00:01:34:07

specific, physical residence (i.e., by matching up the address, you can then find the desired physical location point, as shown in Figure 11.3a). For example, let's suppose that a time-encoded analog multitrack tape begins at time 00:01:00:00, ends at 00:28:19:00 and contains a specific cue point (such as a glass shattering) that begins at 00:12:53:19 (Figure 11.3b). By monitoring the time code readout, it's a simple matter to locate the precise position that corresponds to the cue point on the tape and then perform whatever function is necessary, such as inserting an effect into the sound track at that specific point … CRAAAASH!

Master/Slave Relationship

In light of the "Black Lives Matter" movement, many of us in the technical side of the music production community have struggled with the problematic terms of master/slave that are used within synchronisation. I've even thought about changing the terms myself and letting the chips fall where they may. However, it has been brought to my attention that these terms fall under the MMA Technical Standards Board (as well as from the folks at SMPTE).

These terms are currently under consideration for being changed. Personally, I also struggle with the negative connotations vs. the concept that these terms refer to inanimate devices and not people.

For now, I've decided to follow the current uses of these long-standing terms. For now … all I can say is that "Black Lives Matter!"

FIGURE 11.2
Readout of a SMPTE time code address in HH:MM:SS:FF.

FIGURE 11.3
Location of relative addresses: (a) postal address analogy; (b) time code addresses and a cue point on longitudinal tape.

(a)

(b)

your address
your town
your country

event (BOOM!)

00:28:19:00   00:12:53:19   00:01:00:00

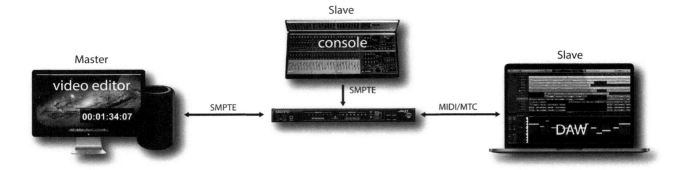

Master
Slave
Slave

SMPTE   SMPTE   MIDI/MTC

Since synchronization is based on the timing relationship between two or more devices, it follows that the logical way to achieve sync is to have one or more devices known as slaves follow the relative movements of a single transport or device known as the master. The basic rule to keep in mind is that there can be only one master in a connected system; however, any number of slaves can be set to follow the relative movements of a master transport or device (Figure 11.4).

Generally, the rule for deciding which device will be the master in a production system (during the pre-planning phase) can best be determined by asking a few questions:

- What type of media is the master time code media recorded on?
- Which device will provide the most stable timing reference?
- Which device will most easily and cost-effectively serve as the master?

If the master comes to you from an outside source, asking lots of questions about the source specs will most likely solve many of your problems. If the project is in-house and you have total say in the matter, you might want to research your options more fully, to make the best choice for your own facility. The following sections can help give you insights into which devices will best serve as the master within a particular system.

Time Code Word

The total of all time-encoded information that's encoded within each audio or video sync frame is known as a time code word. Each word is divided into 80 equal segments, which are numbered consecutively from 0 to 79. One word covers an entire audio or video frame, such that for every frame there is a unique and corresponding time code address. Address information is contained in the digital word as a series of bits that are made up of binary 1s and 0s.

In the case of an analog, a SMPTE signal is electronically encoded in the form of a modulated square wave. This method of encoding information is known as bi-phase modulation. Using this code type, a voltage or bit transition in the middle of a half-cycle of a square wave represents a bit value of 1, while no transition within this same period signifies a bit value of 0 (Figure 11.5). The most important feature about this system is that detection relies on shifts within

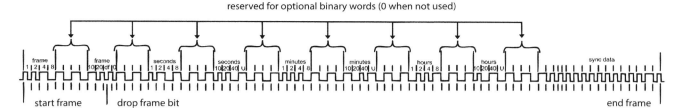

reserved for optional binary words (0 when not used)

frame 1|2|4|8   frame 10|20|df|0   seconds 1|2|4|8   seconds 10|20|40|U   minutes 1|2|4|8   minutes 10|20|40|U   hours 1|2|4|8   hours 10|20|40|U   sync data

start frame   drop frame bit   end frame

the pulse and not on the pulse's polarity or direction. Consequently, time code can be read in either the forward or reverse play mode, as well as at fast or slow shuttle speeds.

FIGURE 11.5
Bi-phase representation of a SMPTE time code word.

Time Code Frame Standards

In productions using time code, it's important that the readout display be directly related to the actual elapsed time of a program, particularly when dealing with the exacting time requirements of broadcasting. Due to historical and technical differences between countries, time code frame rates may vary from one medium, production house or region of origin to another. The following frame rates are available:

- 30 fr/sec (monochrome US video): in the case of a black-and-white (monochrome) video signal, a rate of exactly 30 frames per second (fr/sec) is used. If this rate (often referred to as non-drop code) is used on a black-and-white program, the time code display, program length and actual clock-on-the-wall time would all be in agreement.

- 29.97 fr/sec (drop-frame time code for color NTSC video): the simplicity of 30 fr/sec was eliminated, however, when the National Television Standards Committee (NTSC) set the frame rate for the color video signal in the United States and Japan at 29.97 fr/sec. Thus, if a time code reader that's set up to read the monochrome rate of 30 fr/sec were used to read a color program, the time code readout would pick up an extra 0.03 frame for every second that passes. Over the duration of an hour, the time code readout would differ from the actual elapsed time by a total of 108 frames (or 3.6 seconds). To correct for this difference and bring the time code readout and the actual elapsed time back into agreement, a series of frame adjustments was introduced into the code. Because the goal is to drop 108 frames over the course of an hour, the code used for color has come to be known as drop-frame code. In this system, two frame counts for every minute of operation are omitted from the code, with the exception of minutes 00, 10, 20, 30, 40 and 50. This has the effect of adjusting the frame count, so that it agrees with the actual elapsed duration of a program.

- 29.97 fr/sec (non-drop-frame code): in addition to the color 29.97 drop-frame code, a 29.97 non-drop-frame color standard can also be found in video production. When using non-drop time code, the frame count will always advance one count per frame, without any drops. As you might expect, this mode will result in a disagreement between the frame count and the actual clock-on-the-wall time over the course of the program. Non-drop, however, has the distinct advantage of easing the time calculations that are often required in the video-editing process (because no frame compensations need to be taken into account).

- 25 fr/sec EBU (standard rate for PAL video): another frame rate format that's used throughout Europe is the European Broadcast Union (EBU) time code. EBU utilizes SMPTE's 80-bit code word but differs in that it uses a 25 fr/sec frame rate. Because both monochrome and color video EBU signals run at exactly 25 fr/sec, an EBU drop-frame code isn't necessary.
- 24 fr/sec (standard rate for film work): the medium of film differs from all of these in that it makes use of an SMPTE time code format that runs at 24 fr/sec.

From the above, it's easy to understand why confusion often exists as to which frame rate should be used on a project. Basically, if you are working on an in-house project that doesn't incorporate time-encoded material that comes from the outside world, you should choose a rate that both makes sense for you and is likely to be compatible with an outside facility (should the need arise).

For example, electronic musicians who are working in-house in the US will often choose to work at 30 fr/sec. Those in Europe have it easy, because on that continent 25 fr/sec is the logical choice for all music and video productions. On the other hand, those who work with projects that come through the door from other production houses will need to take special care to reference their time-code rates to those used by the originating media house. This can't be stressed enough: if care isn't taken to keep your time code references at the proper rate and relative address times (while keeping degradation to a minimum from one generation to the next), the various media might have trouble syncing up when it comes time to put the final master together … and that could spell BIG trouble.

Time Code within Digital Media Production

Given that SMPTE exists in a digitally encoded data form, current-day digital professional media devices are able to accept and communicate SMPTE directly without too much trouble. Professional camera, film, controllers and editing systems are able to directly chain and synchronize SMPTE using a multitude of complicated, yet standardized methods that make use of both digital- and analog-style time code data streams.

Of course, there are a wide range of approaches that can be taken when media devices (cameras, video editing software and field audio recorders) are to be synchronized together. These can range from "shoots" that make use of multiple cameras and separate field recorders, which are "locked" to a single time code source on a set … all the way down to a simple camera and digital hand recorder, with audio that can be manually synced up within the digital editor, without the use of time code at all. The types of equipment and the ways that they deal with the technology of sync are ever-changing. Therefore it's important to keep abreast of current technology, read the manuals (about how connections and settings can best be made) and dive into the study of visual media production.

BROADCAST WAVE FILE FORMAT

Although digital media devices and software are able to import, convert and communicate using the language of SMPTE time code (in all its various flavors), the folks at the EBU (European Broadcast Union) saw the need to create a universal audio file format that would include time code data within all forms of audio and visual media production. The result was the *Broadcast Wave Format* (BWF). Broadcast Wave is in most ways completely compatible with its Microsoft Wave counterpart, with the exception that it is able to embed metadata (information about the recorded content ... photo, take#, date, technical data, etc.) as well as SMPTE time code address data. The inclusion of such important content and time code information means that the time-related information will actually be imbedded within the file itself, allowing sound files that are imported into a video or audio editor to automatically snap to their appropriate time code position. Obviously, Broadcast Wave can be a huge time saver within the production and post-production process.

SYNC IN THE PRE-MIDI ERA

Before the MIDI specification was implemented, electronic instruments and devices used other types of synchronization methods. Although sync between these non-MIDI and MIDI instruments was a source of mild to major aggravation, a number of these older devices can still be found within MIDI setups because of their distinctive and wonderful sounds.

Click Sync

Click sync or click track refers to the metronomic audio clicks that are generated to communicate tempo. These are produced once per beat or once per several beats (as occurs in cut time or compound meters). Often, a click or metronome is designed into a MIDI interface or sequencing software to produce an audible tone or to trigger MIDI instrument notes that can be used as a tempo guide. These guide clicks help a musician keep in tempo with a sequenced composition. Certain sync boxes and older drum machines can sync a sequence to a live or recorded click track. They can do this by determining the beat based on the tempo of the clicks and then output a MIDI start message (once a sufficient number of click pulses has been received for tempo calculation). A MIDI stop message might be transmitted by such a device whenever more than two clicks have been missed or whenever the tempo falls below 30 beats/minute. Note that this sync method doesn't work well with rapid tempo changes, because chase resolutions are limited to one click per beat (1/24th the resolution of MIDI clock). Thus, it's best to use a click source that is relatively constant in tempo.

TTL and DIN Sync

One of the most common ways to lock sequencers, drum machines and instruments together, before the adoption of MIDI, was TTL 5-volt sync. In this system, a musical beat is divided into a specific number of clock pulses per quarter note (PPQN), which varies from device to device; for example, DIN sync (a form of TTL sync, which is named after the now famous 5-pin DIN connector) is transmitted at a rate of 24 PPQN. TTL can be transmitted in either one of two ways. The first and simplest way uses a single conductor that passes a 5-volt clock signal. Quite simply, once the clock pulses are received by a slave device, it will start playing and synchronize to the incoming clock rate. Should these pulses stop, the devices will also stop and wait for the clock to again resume. The second method uses two conductors, both of which transmit 5-volt transitions; however, with this system, one line is used to constantly transmit timing information, while the other is used for start/stop information.

FSK

As technology developed, musicians discovered that it was possible to lock instruments that used TTL 5-volt sync to an analog tape recorder. This was done by recording a master sync square-wave pulse onto tape (Figure 11.6a). Since the most common pulse in use at the time was 24 and 48 PPQN, the recorded square wave consisted of an alternating 24- or 48-Hz signal. Although this system worked, it wasn't without its difficulties, because the synchronized devices relied on the integrity of the square wave's sharp transition edges to provide a clean clock reference. Because tape is notoriously bad at reproducing a square wave (Figure 11.6b), the poor frequency response and reduced reliability at low frequencies meant that a better system for synchronizing MIDI-to-tape had to be found. The initial answer was in frequency shift keying, better known as FSK.

FSK works in much the same way as the TTL sync track. However, instead of recording a low-frequency square wave onto tape, FSK uses two, high-frequency square-wave signals for marking clock transitions (Figure 11.6c). In the case of the MPU-401/compatible interface, these two frequencies are 1.25 and 2.5 kHz; the rate at which these pitches alternate determines the master timing clock to which all slaved devices are synched. These devices are able to detect a change in modulation, convert these into a clock pulse and advance their own clocks – hopefully in sync.

FIGURE 11.6
TTL and FSK sync waveforms: (a) original TTL square-wave pulse; (b) playback of TTL sync pulse from tape; (c) modulated FSK sync pulse.

Unlike most other forms of sync lock, FSK triggers and plays the sequence relative to the initial clock pulse that's recorded onto tape. As such, the sequence MUST be positioned at its beginning point and the tape MUST be cued to a

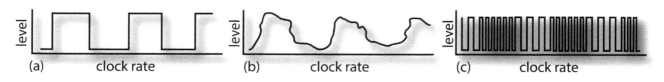

(a) level — clock rate   (b) level — clock rate   (c) level — clock rate point before the beginning of the song. Once the initial sync pulse is received, the sequencer will begin playback. Should a mistake happen, you'll need to recue the song back to its beginning point and start again.

MIDI-BASED SYNC

From the previous section, you can see that a fair degree of math, knowledge of music tempo and time signatures and a large dose of good luck was required to get electronic instruments and media devices to sync up with each other. As a result, it became clear to those in the electronic music/instrument industry that a more reliable form of sync was needed.

MIDI Sync

Before we begin talking about a mechanism for providing sync between MIDI and various other media devices (DAWs, video recorders, digital consoles and the like), I think it's important that we take a look literally into the pulse of the MIDI 1.0 data stream itself. Unlike time code, these MIDI-based communication protocols are used for communicating timing elements between connected MIDI instruments and devices.

MIDI SYSTEM REAL-TIME MESSAGES

As we saw in Chapter 2, MIDI voice messages are specific to a MIDI channel and will be instructed to communicate with a single instrument that's connected in the MIDI chain. On the other hand, MIDI system real-time messages are not specific to a channel and thus communicate information to "all" devices that can respond to these messages within the connected network.

Before we begin, let's dispel the idea that MIDI sync has anything to do with time code or any other kind of time-stamped code that can sync up with other media devices. Instead, MIDI sync is the internal timing pulse that's imbedded within the MIDI data stream itself.

Although this is rarely used within a standard MIDI setup, it's interesting to note that MIDI has a built-in (and often transparent) protocol for synchronizing all of the tempo and timing elements of each MIDI device in a system to a master clock. This functions by transmitting real-time messages to the various devices through standard MIDI cables, USB or other internal connection paths. Although these relationships are often automatically defined within a system setup, one MIDI device must be designated as the master device in order to provide the timing information to which all other slaved devices are locked. MIDI real-time messages consist of four basic types that are each 1 byte in length:

- Timing clock – a clock timing that's transmitted to all devices in the MIDI system which uses divisions known as *pulses per quarter note* (*ppq*). This method is used to improve the system's timing resolution and simplify timing when working in nonstandard meters (e.g., 3/8, 5/16, 5/32). By default, the time signature is 4/4 and the tempo is 120 beats per minute. The ppq is stated in the last word of information (the last two bytes) of the header chunk that appears at the beginning of the file. It could be a low number such as its standard 24 ppq or 96 (both are often enough resolution for most music productions), or it could be a larger number, such as 480 for higher resolution.

- Start – upon receipt of a timing clock message, the start command instructs all connected devices to begin playing from the beginning of their internal sequences. Should a program be in mid-sequence, the start command repositions the sequence back to its beginning, at which point it begins to play.
- Stop – upon the transmission of a MIDI stop command, all devices in the system stop at their current positions and wait for a message to follow.
- Continue – following the receipt of a MIDI stop command, a MIDI continue message instructs all instruments and devices to resume playing from the precise point at which the sequence was stopped. Certain older MIDI devices (most notably drum machines) aren't capable of sending or responding to continue commands. In such a case, the user must either restart the sequence from its beginning or manually position the device to the correct measure.

INTERNAL/EXTERNAL SYNC USING SYSTEM REAL-TIME MESSAGES

I'm totally aware that in this chapter, we're moving between things that are relevant to those that are far less relevant to your everyday workflow. This section is one of those that can seem totally useless in practice … until you REALLY need it!

Here, we are largely talking about hardware instruments that have their own sets of sync/timing elements that are important to a song. In my own case, this relates to hardware groove machines or complex wave-dance patterns that rhythmically evolve over time.

Let's assume that we have a music session up that includes a passage that requires the use of a hardware instrument that has just the right evolving sound-patch that includes a complex percussion groove within its overall sound. Let's say that the song is crafted at a tempo of 100 bpm. So, we take the time to set the tempo of the groove instrument to also be 100 bpm … set up all of our proper routings and press "play". Of course the rhythms will play in perfect sync to the song, right? … Well, quite often, it won't be in rhythmic sync at all. Even though the notes are right and the tempos are right, it's entirely possible that the complex rhythms could be slightly or even wildly out of sync with the actual tempo of the song. This is because the synth is using its own internal MIDI timing clock as a timing reference. The idea here is to turn off "internal sync" (where the device uses its own clock and tempo information for timing) and to set it to "external sync" (whereby it essentially becomes a slave and waits for timing clock signals to come in over MIDI), so that the instrument will get its timing information from the designated master MIDI clock (which is usually your DAW/sequencer). Using this approach, you stand a much better chance that the groove will fall precisely within the original session's tempo timing.

Another example where setting a groove instrument to "external" can be a huge time saver is if you are trying to record individual grooves into DAW for archiving or for use in later sessions. By setting the instrument to "external" you can then set the DAW to record only 4 bars of each groove part (for example). Once you begin playing the sequence, the DAW will instruct the groove box to begin playing in tempo and at the right starting point (possibly set 4 bars before the DAW record-in point is set). In this way the groove will begin playing, after 4 bars the DAW track will switch into record and after the 4-bar duration, it'll drop out of record … and you have your groove in perfect time, ideally with little muss and fuss.

MIDI TIME CODE

MIDI time code (MTC) was developed to allow electronic musicians, project studios, video facilities and virtually all other production environments to cost-effectively and easily translate time code into time-stamped messages that can be transmitted over MIDI data lines. Created by Chris Meyer and Evan Brooks, MIDI time code allows SMPTE-based time code to be distributed throughout the MIDI chain to devices or instruments that are capable of synchronizing to and executing MTC commands. MIDI time code is an extension of the MIDI standard, making use of existing SysEx message types that were either previously undefined or were being used for other, non-conflicting purposes.

Since most modern recording systems include MIDI in their design, there's often no need for external hardware when making direct connections. Simply chain the MIDI data lines from the master to the appropriate slaves within the system (via physical cables, USB or virtual internal routing). Although MTC uses a reasonably small percentage of MIDI's available bandwidth (about 7.68% at 30 fr/sec), it's customary (but not necessary) to separate these lines from those that are communicating performance data when using physical MIDI cables. As with conventional SMPTE, only one master can exist within an MTC system, while any number of slaves can be assigned to follow, locate and chase to the master's speed and position. Because MTC is easy to use and is often included free in many system and program designs, this technology has grown to become the most straightforward and commonly used way to lock together such devices as DAWs, external devices and basic analog and video setups.

MIDI Time Code Messages

The MIDI time code format can be divided into two parts:

- Time code
- MIDI cueing

The time code capabilities of MTC are relatively straightforward and allow devices to be synchronously locked or triggered to SMPTE time code. MIDI cueing is a format that informs a MIDI device of an upcoming event that's to be performed at a specific time (such as load, play, stop, punch-in/out, reset). This protocol envisions the use of intelligent MIDI devices that can prepare for a specific event in advance and then execute the command on cue.

MIDI time code is made up of three message types:

- **Quarter-frame messages**: these are transmitted only while the system is running in real or variable speed time, in either forward or reverse direction. True to its name, four quarter-frame messages are generated for each time code frame. Since eight quarter-frame messages are required to encode a full SMPTE address (in hours, minutes, seconds and frames: 00:00:00:00), the complete SMPTE address time is updated once every two frames (in other words, MIDI time code actually has half the resolution accuracy of its SMPTE time code counterpart). Each quarter-frame message contains 2 bytes. The first byte is F1, the quarter-frame common header; the second byte contains a nibble (4bits) that represents the message number (0 through 7) and a nibble for encoding the time field digit.

- **Full messages**: quarter-frame messages are not sent in the fast-forward, rewind or locate modes, because this would unnecessarily clog a MIDI data line. When the system is in any of these shuttle modes, a full message is used to encode a complete time code address. After a fast shuttle mode is entered, the system generates a full address message and then places itself in a pause mode until the time-encoded slaves have located to the correct position. Once playback has resumed, MTC will again begin sending incremental quarter-frame messages.

- **MIDI cueing messages**: MIDI cueing messages are designed to address individual devices or programs within a system. These 13-bit messages can be used to compile a cue or edit decision list, which in turn instructs one or more devices to play, punch in, load, stop and so on, at a specific time. Each instruction within a cueing message contains a unique number, time, name, type and space for additional information. At the present time, only a small percentage of the possible 128 cueing event types have been defined.

SMPTE/MTC Conversion

Although MIDI time code connections can be directly made between compatible MIDI devices, an SMPTE-to-MIDI converter is required to read incoming LTC SMPTE time code and then convert it into MIDI time code (and vice versa) for other device types. These conversion systems are available as a stand-alone device or as an integrated part of an audio interface or multiport MIDI interface/patch bay/synchronizer system (Figure 11.7).

SYNC IN THE DIGITAL AGE

Obviously, in this age of the digital production workstation, it's often a simple matter to edit a piece of video, audio or MIDI media and import it into a DAW, whereby it can be locked in time to time code, slipped in relative time

without any difficulty or mangled in any way possible. In short, whenever media is imported into a digital timeline that can accept multiple media forms, practically anything is possible and recallable at any time.

FIGURE 11.7
SMPTE time code can often be generated throughout a production system, possibly as either LTC or as MTC via a capable MIDI or audio interface.

DAW Support for Video, Picture and Other Media Sync

Most modern DAWs include support for displaying a video track within a session, both as a video window that can be displayed on the monitor desktop and in the form of a video thumbnail track that will often appear in the track view as a linear guide track. Both of these provide important visual cues for tracking live music, sequencing MIDI tracks and accurately placing automation moves and effects (SFX) at specific hit-points within the scene (Figure 11.8). This feature allows audio to be built up within a DAW environment without the need to sync to an external device at all. It's easily conceivable that through the use of recorded tracks, software instruments and internal mixing capabilities, tracks could be built up, spotted and mixed – all inside the box.

Routing Time Code to and from Your Computer

From a connections standpoint, most DAW, MIDI and audio application software packages are flexible enough to let you choose from any number of available sync sources (whether connected to a hardware port, MIDI interface port or virtual sync driver). All you have to do is assign all of the slaves within the system to the device driver that's generating the system's master code. In most cases, where the digital audio and MIDI sequencing applications are operating

FIGURE 11.8
Most high-end DAW systems are capable of importing a video file directly into the project session window.

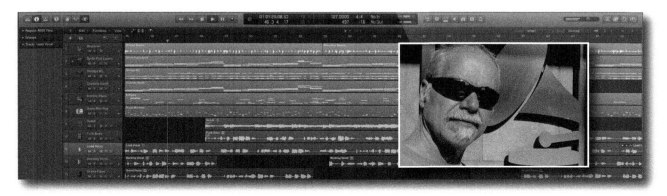

within the same computer, it's best to have your DAW or editor generate the master code for the system.

From time to time, you might run into an application or editor that's unable to generate time code in any form. When faced with such an all-slave software environment, the easiest solution is to use a multiport MIDI interface that includes a software applet for generating time code. In such a situation, all you need to do is to select the interface's sync driver as your sync source for all slave applications. Pressing the Generate SMPTE button in the interface's application window or from its front panel will lock the software to the generated code, beginning at 00:00:00:00 or at any specified offset address.

KEEPING OUT OF TROUBLE

Here are a few guidelines that can help save your butt when using SMPTE and other time code translations during a project:

- When in doubt about frame rates, special requirements or anything else, for that matter ... ask! You (and your client) will be glad you did.
- Fully document your time code settings, offsets, start times, etc.
- If the project isn't to be used completely in-house, ask the producer what the proper frame rate or any other sync parameters should be. Never assume or guess it.
- When beginning a new session (when using a tape-based device), always stripe the master contiguously from the beginning to end before the session begins. It never hurts to stripe an extra tape, just in case.
- Start generating new code at a point 1 to 2 minutes before 01:00:00:00 or 00:01:00:00 (to allow for a pre-roll ... especially when an analog device is involved). If the project isn't to be used in-house, ask the producer what the start times should be. Don't assume or guess it.
- Never dub (copy) an analog copy of time code directly. Always make a refreshed (jam synched) copy of the original time code (from an analog master) before the session begins.
- Disable noise reduction on analog audio tracks (on both audio and video decks).

In closing, I'd like to point out that synchronization can be a simple procedure or it can be a complex one, depending on your experience, the situation and the type of equipment that's involved. A number of books and articles have been written on this subject. If you're serious about production, I suggest that you do your best to keep up on the topic. Although the fundamentals stay the same, new technologies and techniques are constantly emerging that require that you stay on your toes. As always, the best way to learn is simply by reading about it and then jumping in and doing it.

CHAPTER 12

Multimedia and the Web

It's no secret that modern-day computers, smart phones, game-stations and even smart televisions have gotten faster, sleeker, more touchable and sexier in their overall design. In addition to their ability to act as a multifunctional production workhorse, one of the crowning achievements of modern work and entertainment devices is their networking and media integration, which has come to be universally known by the household buzzword *multimedia*.

This combination of working and playing with multimedia has found its way into modern media and computer culture through the use of various hardware and software systems which combine in a multitasking environment to bring you an experience that seamlessly involves such media types as:

- Audio and music
- Video and video streaming
- Graphics and gaming
- Musical instrument digital interface (MIDI)
- Text and communication

The obvious reason for creating and integrating these media types is the human desire to share and communicate one's experiences with others. This has been done for centuries in the form of books and, in relatively more recent decades, through movies and television. Obviously, in the here and now, the powerful and versatile presence of the WWW can be placed near the top of this communications list. Nothing allows individuals and corporate entities to reach millions (or billions) so easily. Perhaps most importantly, the Web is a multimedia experience that each individual can manipulate, learn from and even respond to in an interactive fashion. It has indeed unlocked the potential for experiencing events and information in a way that makes each of us a participant, and not just a passive spectator. To me, this is the true revolution that's occurring at the dawn of the 21st century!

THE MULTIMEDIA ENVIRONMENT

Although much of recording and music production has matured into a relatively stable industry, the Web, multimedia and the music industry itself are in a full-speed-ahead tailspin of change. With online social media, on-demand video and audio streaming, network communications, computer gaming and hundreds of other media option entering onto the marketplace on a weekly basis … it's no wonder that things are changing fast!

As with all things tech, I would say that the most important word in multimedia technology today is "integration". In fact, the perfect example of multimedia today is in your pocket or purse. Your cell phone (handy, mobile or whatever you call it) is a marvel of multimedia technology that can:

- Keep you in touch with friends
- Surf the Web to find the best restaurant in the area
- Connect to a Web-based and/or actual GPS to keep you from getting lost
- Play or stream your favorite music
- Let you watch the latest movie or YouTube video
- Take pictures
- Light your way to the bathroom late at night or remotely turn your home's lights on or off

… Of course, that's just the short list. What more could you ever want from a mobile on-the-go device? I don't know … but we're surely to find out in the not-too-distant-future, and, because of these everyday tools, we've come to expect the same or similar experience from other such media devices as:

- Computers
- Television
- The car

The Computer

Obviously, the tool that started the multimedia revolution is the computer. The fact that it is a lean, mean multitasking machine makes it ideal for delivering all of the media that we want, all of the time. From multimedia, to gaming, to music, to video … if you have a good Internet connection … you can pretty much hold the world in your hands (or at least your lap).

Television and the Home Theater

As you might expect, newer generations of video and home theater systems incorporate more and more options for offering up "a rich multimedia experience". Newer "smart" TVs are able to directly connect to the Web, allowing us to stream our favorite movies or listen to Internet radio "on demand".

So, what holds all of these various media devices together? Media and data distribution and transmission formats!

Delivery Media

Although media data can be stored and/or transmitted on a wide range of storage devices, the most commonly found delivery media at the time of this writing are:

- Shared networks
- The Web
- Physical media

Networking

At its most basic level, a shared data *network* is a collection of computers and other hardware devices that are connected by protected data protocol links that allow for the communication of shared resources, program apps and data. The most well-known network communications protocols are the Ethernet (a standard for creating local area networks), and the Internet protocol suite (otherwise known as the WWW).

Within media production, a local area network (or LAN) is a powerful tool that allows multiple computers to be linked within a larger production facility. For example, a central server (a dedicated data delivery and shared storage device) can be used in a facility to store large amounts of data throughout a facility. Simpler LAN connections could also be used in a project studio to link various computers, so as to share media and data backup in a simple and cost-effective manner. For example, a single, remote computer within a connected facility could be used to store and share the large amounts of data that are required for video, music and sample library production … and then backup all of these data in a redundant array of independent disks (RAID) system, allowing the data and backups to be duplicated and stored on multiple hard drives (thereby reducing the chance of system data loss).

In short, it's always a good idea to become familiar with the strength, protection and power that a properly designed network can offer a production facility, no matter how big it is.

The Web

One of the most powerful aspects of multimedia is its ability to communicate experiences either to another individual or to the masses. For this, you need a very large network connection. The largest and most common network of all is the Internet (World Wide Web). Here's the basic gist of how this beast works:

- The Internet (Figure 12.1) can be thought of as a communications network that allows your computer (or connected network) to be connected to an Internet Service Provider (ISP) server (a specialized computer or cluster of ISP computers that are designed to handle, pass and route data between other network user connections).
- These ISPs are then connected (through specialized high-speed connections) to a series of network access points (NAPs), which essentially form the connected infrastructure of the World Wide Web (WWW).

mac   server   WWW   server   pc

FIGURE 12.1
The Internet works by communicating requests and data from a user's computer to connected servers that are connected to other network access points around the world, which are likewise connected to other users' computers.

Therefore, in its most basic form, the Web can be simply thought of as a unified array of connected networks.

Internet browsers transmit and receive information on the Web via a Uniform Resource Locator (URL) address. This address is then broken down into three parts: the protocol (e.g., http:// or https://), the server name (e.g., www.modrec.com) and the requested page or file name (e.g., www.modrec.com/index.htm). The connected server is able to translate the server name into a specific Internet provider (IP) address, which is then used to connect your computer to the desired server, after which the requests to receive or send data are communicated and the information is passed to your computer.

Email works in a similar data transfer fashion, with the exception that an email isn't sent to or requested from a specific server; rather, it's communicated through a worldwide server network from one specific email address (e.g., myname@myprovider.com) directly to a destination email address (e.g., yourname@yourprovider.com).

The Cloud

One of the more current buzz terms on the Web is "The Cloud" or "Cloud Computing". Put simply, storing data on the cloud refers to data that are stored on a remote server system or Web-connected drive system. Most commonly, that server would be operated and maintained by a company that would store your data at a cost (although many services allow limited amounts of your personal data to be stored for free). For example, cloud storage companies can:

- Store backup media online in a manual or automated way to keep your files secure.
- Store huge amounts of uploadable video data online in a social media context (e.g., YouTube).
- Store uploadable audio data online in a social media context for promoting artists and DJs (i.e., Bandcamp, ReverbNation, SoundCloud).
- Store program-related data online, reducing the need for installing programs directly onto your computer.
- Store application and program data online, allowing the user to "subscribe" to the use of their programs for a monthly or yearly fee.

PHYSICAL MEDIA

Although online data distribution and management is increasingly becoming the primary way to get information from the distributor to the consumer, physical media (you know, the kind that we can hold in our hands) still has a very important role in delivering media to the consumer masses. Beyond the fun stuff – like vinyl records – the most common physical media formats are CDs, DVDs and Blu-ray discs.

The CD

Of course, one of the first and most important developments in the mass marketing and distribution of large amounts of digital media was the compact disc (CD), both in the form of the CD-Audio and the CD-ROM. As most are aware, the CD-Audio disc is capable of storing up to 74 minutes of audio at a rate of 16 bits/44.1 kHz. Its close optical cousin, the CD-ROM, is most often capable of storing 700 MB of graphics, video, digital audio, MIDI, text and raw data. Consequently, these pre-manufactured and user-encoded media are still widely used to store large amounts of music, text, video, graphics, etc. … in part due to the fact that you can hold it in your hand and store it away in a safe place.

Table 12.1 details the various CD standards (often affectionately called the "rainbow book") that are currently in use.

It's important to note that Red Book CDs (audio CDs) are capable of encoding small amounts of user-data that can be used to encode imbedded metadata (user information). This metadata (called CD-Text) allows the media creator to enter and encode information such as artist, title, song number, track artist/title, etc. … all within the disc itself. Since most CD players, computers and media players are capable of reading this information, it's always a good idea to provide the listener with as much information as possible.

Another system for identifying CD and disc-related data is provided by Gracenote (formerly Compact Disc Data Base [CDDB]). In short, Gracenote maintains and licenses an Internet database that contains CD info, text and images. Many media devices that are connected to the Web are able to access this database and display the information on your player or device.

The DVD

Similar to their cousin, the DVD (which, after a great deal of industry deliberation, simply stands for "DVD") can contain any form of data. These discs are capable of storing up to 4.7 gigabytes (GB) within a single-sided disc or 8.5 GB on a double-layered disc. This capacity makes the DVD a good delivery medium for encoding video (generally in the MPEG-2 encoding format), data-intensive games, DVD-ROM titles and program installation discs. The increased demand for multimedia games, educational products, etc., has spawned the

Table 12.1	CD Format Standards
Format	**Description**
Red Book	Audio-only standard; also called Compact Disc Audio (CD-A)
Yellow Book	Data-only format; used to write/read CD-ROM data
Green Book	Compact Disc Interactive (CD-I) format; never gained mass popularity
Orange Book	Compact Disc Recordable (CD-R) format
White Book	Video Compact Disc (VCD) format for encoding CD-A audio and MPEG-1 or MPEG-2 video data; used for home video and karaoke
Blue Book	Enhanced Music CD format (also known as CD Extra or CD+) can contain both CD-A and data
ISO-9660	A data file format that's used for encoding and reading data from CDs of all types across platforms
Joliet	Extension of the ISO-9660 format that allows for up to 64 characters in its file name (as opposed to the 8 file + 3 extension characters allowed by MS-DOS)
Romeo	Extension of the ISO-9660 format that allows for up to 128 characters in the file name
Rock Ridge	Unix-style extension of the ISO-9660 format that allows for long file names
CD-ROM/XA	Allows for extended usage for the CD-ROM format – Mode-1 is strictly Yellow Book, while Mode-2 Form-1 includes error correction and Mode-2 Form-2 doesn't allow for error correction; often used for audio and video data
CD-RFS	Incremental packet writing system from Sony that allows data to be written and rewritten to a CD or CD-RW (in a way that appears to the user much like the writing/retrieval of data from a hard drive)
CD-UDF	Universal Disc Format (UDF) is an open incremental packet writing system that allows data to be written and rewritten to a CD or CD-RW (in a way that appears to the user much like the writing/retrieval of data from a hard drive) according to the ISO-13346 standard
HDCD	The High-Definition Compatible Digital system adds 6 dB of gain to a Red Book CD (when played back on an HDCD-compatible player) through the use of a special companion mastering technique
Macintosh HFS	An Apple file system that supports up to 31 characters in a file name; includes a data fork and a resource fork that identify which application should be used to open the file

computer-related industry of CD and DVD-ROM authoring. The term *authoring* refers to the creative, design and programming aspects of putting together a CD or DVD project. At its most basic level, a project can be authored, mastered and burned to disc from a single commercial authoring program. Whenever the stakes are higher, trained professionals and expensive systems are often called in to assemble, master and produce the final disc for mass duplication and packaging. Table 12.2 details the various DVD video/audio formats that are currently in use.

Table 12.2	DVD Video/Audio Formats					
Format	**Sample Rate (kHz)**	**Bit Rate**	**Bit/s**	**Ch**	**Common Format**	**Compression**
PCM	48, 96	16, 20, 24	Up to 6.144 Mbps	1 to 8	48 kHz, 16 bit	None
AC3	48	16, 20, 24	64 to 448 kbps	1 to 6.1	192 kbps, stereo	AC3 and 384 kbps, 448 kbps
DTS	48, 96	16, 20, 24	64 to 1,536 kbps	1 to 7.1	377 or 754 kbps for stereo and 754.5 or 1509.25 kbps for 5.1	DTS coherent acoustics
MPEG-2	48	16, 20	32 to 912 kbps	1 to 7.1	Seldom used	MPEG
MPEG-1	48	16, 20	384 kbps	2	Seldom used	MPEG
SDDS	48	16	Up to 1,289 kbps	5.1, 7.1	Seldom used	ATRAC

Blu-Ray

Although similar in size to the CD and DVD, a Blu-ray disc can store up to 25 GB of media-related data onto each data layer (50 GB for a dual-layer disc). In addition to most of the standard video formats that are commonly encoded onto a DVD, the Blu-ray format can playback both compressed and non-compressed PCM audio in a multi-channel, high-resolution environment.

The Flash Card and Memory USB Stick

In our on-the-go world, another useful media device is the flash memory card. More specifically, the SD (secure digital) card typically ranges in size up to 128 Gb in capacity and comes in various physical sizes. These media cards can be used for storing audio, video, photo, app and any other type of digital info that can be used with your laptop, phone, car player, Blu-ray player … you name it!

MEDIA DELIVERY FORMATS

Now that we've taken a look at the various delivery media, the next most important aspect of delivering the multimedia experience rests with the data distribution formats (the coding and technical aspects of data delivery) themselves.

When creating content for the various media systems, it's extremely important that the media format and bandwidth be matched with the requirements of the content delivery system that's being used. In other words, it's always smart to maximize the efficiency of the message (media format and required bandwidth) to match (and not alienate) your intended audience. The following section outlines many standard and/or popular formats for delivering media to a target audience.

Uncompressed Sound File Formats

Digital audio is obviously a component that adds greatly to the multimedia experience. It can augment a presentation by adding a dramatic music soundtrack, help us to communicate through speech or give realism to a soundtrack by adding sound effects. Because of the large amounts of data required to pass video, graphics and audio from a disc, the Internet or other media, the bit- and sample-rate structure of an uncompressed audio file is usually limited compared to that of a professional-quality sound file. At the "lo-fi" range, the generally accepted sound file standard for older multimedia production is either 8-bit or 16-bit audio at a sample rate of 11.025 or 22.050 kHz. This standard came about mostly because older CD drive and processor systems generally couldn't pass the professional rates of 44.1 kHz and higher. With the introduction of faster processing systems and better hardware, these limitations have generally been lifted to include 16/44.1 (16 bit/44.1 kHz), 24/44.1 and as high as 24/192. Obviously there are limitations to communicating uncompressed professional-rate sound files over the Internet or from an optical disc that's also streaming full-motion video. Fortunately, with improvements in codec (encode/decode) techniques, hardware speed and design, the overall sonic and production quality of compressed audio data has greatly improved.

PCM Audio File Formats

Although several formats exist for encoding and storing sound file data, only a few have been universally adopted by the industry. These standardized formats make it easier for files to be exchanged between compatible media devices.

In audio, pulse-code modulation (PCM) is the standard system for encoding, storing and decoding audio. Within a PCM stream, the amplitude of the analog signal is sampled at precise intervals, with each sample being quantized to the nearest value within a range of digital steps. This level (amplitude) is then sampled at precise time intervals (frequency), so as to represent analog audio in a numeric form.

Probably the most common file type is the *Wave (or .wav) format*. Developed for the Microsoft Windows format, this universal file type supports both mono and stereo files at a wide range of uncompressed resolutions and sample rates. Wave files contain PCM-coded audio that follows the Resource Information File Format (RIFF) spec, which allows extra user information to be embedded and saved within the file itself. The newly adopted *Broadcast Wave format*, which has been adopted by the Producers and Engineers Wing (www.grammypro.com/pr oducers-engineers-wing) as the preferred sound file format for DAW production and music archiving, allows for time-code-related positioning information to be directly embedded within the sound file's data stream.

In addition to the .wav format, the Audio Interchange File (AIFF; .aif) format is commonly used to encode digital audio within Apple computers. Like Wave files, AIFF files support mono or stereo, 8-bit, 16-bit and 24-bit audio at a wide range of sample rates – and like Broadcast Wave files, AIFF files can also contain embedded text strings. Table 12.3 details the differences between uncompressed file sizes as they range from the 24-bit/192-kHz rates, all the way down to lo-voice quality 8-bit/10-kHz files.

Table 12.3	Audio Bit Rate and File Sizes				
Sample Rate	**Word Length**	**No. of Channels**	**Date Rate (kbps)**	**MB/min**	**MB/hour**
192	24	2	1,152	69.12	4,147.2
192	24	1	576	34.56	2,073.6
96	32	2	768	46.08	2,764.8
96	32	1	384	23.04	1,382.4
96	24	2	576	34.56	2,073.6
96	24	1	288	17.28	1,036.8
48	32	2	384	23.04	1,382.4
48	32	1	192	11.52	691.2
48	24	2	288	17.28	1,036.8
48	24	1	144	8.64	518.4
48	16	2	192	11.52	691.2
48	16	1	96	5.76	345.6
44.1	32	2	352	21.12	1,267.2
44.1	32	1	176	10.56	633.6
44.1	24	2	264	15.84	950.4
44.1	24	1	132	7.92	475.2
44.1	16	2	176	10.56	633.6
44.1	16	1	88	5.28	316.8
32	16	2	128	7.68	460.8
32	16	1	64	3.84	230.4
22	16	2	88	5.28	316.8
22	16	1	44	2.64	158.4
22	8	1	22	1.32	79.2
11	16	2	44	2.64	158.4
11	16	1	22	1.32	79.2
11	8	1	11	0.66	39.6

DSD Audio

Direct Streaming Digital (DSD) was a joint venture between Sony and Phillips for encoding audio onto the Super Audio CD (SACD). Although the SACD has fallen out of favor (with the Blu-ray format's wide acceptance), the audio format itself survives as a high-resolution audio format.

Unlike PCM, DSD makes use of pulse-density modulation (Figure 12.2) to encode audio. That's to say that it doesn't follow PCM's system of the periodic sampling of audio at a specific rate; rather, the level and change of relative gain levels over time is a result of the density of the bits within the stream. A stream that has all 0s will have no level at that point in time, while one having all 1s will have a maximum voltage level. The density of 1s to 0s will determine the overall change in gain over time at a sampling rate of 2.8224 MHz (or 64 times the 44.1-kHz sample rate), 5.6448 MHz (DSD128) or higher. Currently, only a few DAWS are able to work natively in the DSD modulation code.

Compressed Codec Sound File Formats

As was mentioned earlier, high-quality uncompressed sound files often present severe challenges to media delivery systems that are restricted in terms of bandwidth, download times or memory storage. Although the streaming of audio data from various media and high-bandwidth networks (including the Web) has improved over the years, memory storage space and other bandwidth limitations have led to the popular acceptance of compressed audio data formats known as codecs. These formats can encode audio in a manner that reduces data file size and bandwidth requirements and then decode the information upon playback using a system known as perceptual coding.

Perceptual Coding

The central idea behind *perceptual coding* is the psychoacoustic principle that the human ear will not always be able to hear all of the information that's present in a recording. This is largely due to the fact that louder sounds will often mask sounds that are both lower in level and relatively close in frequency to another louder signal. These perceptual coding schemes take advantage of this masking

FIGURE 12.2
Uncompressed audio coding. (a) PCM encodes the absolute level at that sample period and stores that number within memory. (b) DSD, on the other hand, does not encode the level within samples and words, but encodes the "density" of 1s to 0s within a period of time to determine the signal's level … there is no coding, per se.

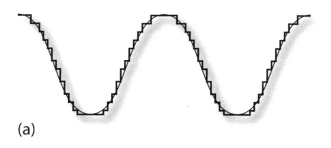

(a)

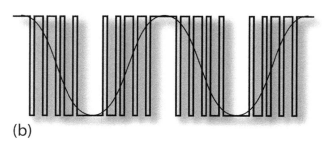

(b)

effect by filtering out noises and sounds that can't be detected by our ears and removes them from the encoded audio-stream.

The perceptual encoding process is said to be "lossy", because once the filtered data has been taken away it can't be replaced or introduced back into the file. For the purposes of audio quality, the amount of perceived data compression reduction can be selected by the user during the encoding process. Higher bandwidth compression rates will remove less data from a stream (resulting in a reduced amount of filtering and higher audio quality), while low bandwidth rates will greatly reduce the data stream (resulting in smaller file sizes, increased filtering, increased artifacts and lower audio quality). The amount of filtering that's to be applied to a file will depend on the intended audio quality and the delivery medium's bandwidth limitations. Due to the lossy character of these encoded files, it's always a good idea to keep a copy of the original, uncompressed sound file in a data archive backup, should changes in content or future technologies occur (never underestimate Murphy's law).

Many of the listed codecs are capable of encoding and decoding audio using a constant bit rate (CBR) and variable bit rate (VBR) structure:

- CBR encoding is designed to work effectively in a streaming scenario where the end user's bandwidth is a consideration. With CBR encoding, the chosen bit rate will remain constant over the course of the file or stream.
- VBR encoding is designed for use when you want to create a downloadable file that has a smaller file size and bit rate without sacrificing sound and video quality. This is carried out by detecting which sections will need the highest bandwidth and adjusting the encode process accordingly. When lower rates will suffice, the encoder adjusts the processing to match the content. Under optimum conditions, you might end up with a VBR-encoded file that has the same quality as a CBR-encoded file, but is only half the file size.

Perceptual coding schemes that are in most common use include:

- MP3
- MP4
- WMA
- AAC
- RealAudio
- FLAC

MP3

MPEG (which is pronounced "M-peg" and stands for the Moving Picture Experts Group; www.mpeg.org) is a standardized format for encoding digital audio into a compressed format for the storage and transmission of various media over the Web. As of this writing, the most popular format is the ISO-MPEG Audio Level-2 Layer-3, commonly referred to as MP3. Developed by the Fraunhofer Institute (www.iis.fraunhofer.de) and Thomson Multimedia in Europe, MP3 has advanced the public awareness and acceptance of compressing and distributing digital audio by creating a codec that can compress audio by a substantial factor while still maintaining quality levels that approach those of a CD (depending on which compression levels are used). Although a wide range of compression rates can be chosen to encode/decode an MP3 file, the most common rate for the general masses is 128 kbps (kilobits per second). Although this rate is definitely "lossy" (containing increased distortion, reduced bandwidth and sideband artifacts), it allows us to literally put thousands of songs on an on-the-go player. Higher rates of 160, 192 and 320 kbps offer higher "near CD sound quality" with the obvious tradeoff being larger file sizes.

Although faster Web connections are commonly used to stream audio in real time, the MP3 format is most often downloaded by the end consumer for storage to disk, disc and SD media for the storage and playback of songs. Once saved, the data can then be transferred to playback devices (such as phones, pads, etc.). In fact, billions of music tracks are currently being downloaded every month on the Internet using MP3, practically every personal computer contains licensed MP3 software and virtually every song has been encoded into this format … it's actually hard to imagine how many players there are out there on the global market. This makes it the Web's most popular audio compression format by far.

MP4

Like MP3, the MPEG-4 (MP4) codec is largely used for streaming video data over the Web or for viewing media over portable devices. MP4 is largely based on Apple's QuickTime "MOV" format and can be used to encode A/V and audio only content over a wide range of bitrate qualities with both stereo and multichannel (surround) capabilities. In addition, this format can employ DRM (copy protection), so as to restrict copying of the downloaded file.

WMA

Developed by Microsoft as their corporate response to MP3, Windows Media Audio (WMA) allows compression rates to encode high-quality audio at low bitrate and file-size settings. Designed for ripping (extracting audio from a CD) and sound file encoding/playback from within the popular Window's Media Player (Figure 12.3), this format has grown and then fallen in general acceptance and popularity. In addition to its high quality at low bit rates, WMA also allows for a wide range of bitrate qualities with both stereo and multichannel (surround) capabilities, while being able to imbed DRM (copy protection), so as to restrict copying of the downloaded file

AAC

Jointly developed by Dolby Labs, Sony, ATT and the Fraunhofer Institute, the Advanced Audio Coding (AAC) scheme is touted as a multichannel-friendly format for secure digital music distribution over the Internet. Stated as having the ability to encode CD-quality audio at lower bit rates than other coding formats, AAC not only is capable of encoding 1, 2 and 5.1 surround sound files but can

also encode up to 48 channels within a single bitstream at bit/sample rates of up to 24/96. This format is also Secure Digital Music Initiative (SDMI)-compliant, allowing copyrighted material to be protected against unauthorized copying and distribution. AAC is the default or standard audio format for YouTube, iPhone, iPod, iPad, iTunes (it's the backbone of Apple's music and audio media distribution), Nintendo DSi, Nintendo 3DS, DivX Plus Web Player and PlayStation 3.

FIGURE 12.3
Window's Media Player.

FLAC

Free Lossless Audio Codec (FLAC) is a format that makes use of a data compression scheme that's capable of reducing an audio file's data size by 40% to 50%, while playing back in a lossless fashion that maintains the sonic integrity of the original stereo and multichannel source audio (up to 8 channels, bit depths of up to 32 bits at sample rates that range to 655.350 kHz). As the name suggests, FLAC is a free, open-source codec that can be used by software developers in a royalty-free fashion.

With the increase in memory storage size and higher download speeds, many enthusiasts in audio are beginning to demand higher playback quality. As such, FLAC is growing in popularity as a medium for playing back high-quality, lossless audio, both in stereo and in various surround formats.

Tagged Metadata

Within most types of multimedia file formats it's possible to embed a wide range of content identifier data directly into the file itself or within a Web-related page or event. This "tagged" data (also known as *metadata*) can identify and provide extensive and extremely important information that relates to the content of the file. For example, let's say that little Sally is looking to find a song that she wants to download from her favorite artist. Now, Sally consumes a lot of music and she goes to iTunes to download that song into her phone. So, how can she find her favorite needle in a digital haystack? By searching for songs under "Mr. Right", the site is able to find several of his latest songs and BOOM ... she's groovin' to the beat ... all thanks to metadata.

Now that she knows the name of the song, she can download it to her player and she's groovin' on the underground or heading to school ... On the flip-side, if the song name hasn't been entered (or was incorrectly entered) into the metadata database, poor Sally's song would literally get lost in the shuffle. Sally would be bummed and the record label/artist would lose the sale.

On another day, let's say that Sally *really* wanted to buy a new song. She could enter her favorite music genre into the field and look through the songs that have been properly tagged with that genre. By clicking the "sounds like" button, songs that have been tagged in the same genre could pop up that might completely flip her out ... BOOM, again ... a sale and (even better) a fan of a new artist is born ... all because the "sounds like" metadata was properly tagged by her new, favorite band. Are you getting the idea that "tagging" a song, project or artist band data with the proper metadata can be a make or break deal in the new digital age?

Metadata in all its media and website glory tells the world who the artist is, the song title, what genre type, etc. It's the gateway to telling the world "Hey, I'm here ... Listen to me!" Metadata can also be extracted from an audio CD using a central music database and then be automatically entered into a music copy (ripping) program (Figure 12.4).

Due to the fact that there is no existing set of rules for filling out metadata, folks who make extensive use of iTunes, Discogs and other music playlist services often go nuts when the tags that are incorrect, conflicting or non-standard. For example, a user might download one album that might go by the artist name of "XYZ Band" and then download another album might have it listed as "The Band XYZ". One would show up in a player under "X" while the same band would also show up under "T" ... and it can get A LOT worse than that. In fact, many people actually go so far as to manually fix their library's metadata to their own liking. The moral of this story is to research your metadata and stick to a single, consistent naming scheme.

FIGURE 12.4
Embedded metadata file tags can be added to a media file via various media libraries, rippers or editors and then viewed by a media player or file manager.

It's worth mentioning here that the Producers and Engineers Wing of the Grammys have been doing extensive work on the subject of metadata (www.grammy.com /credits), so that tags can be more uniform and extensive, allowing producer, engineer and other project-related data to be entered into the official metadata. It is hoped that such credit documentation will help to get royalties to those who legally deserve to be recognized and paid for their services in the here-and-now or in the future (when new payment legislations might be passed).

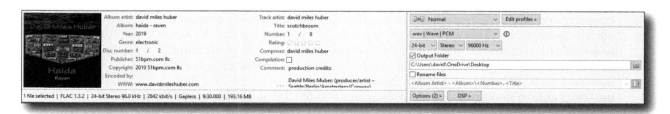

MIDI

One of the unique advantages of MIDI as it applies to multimedia is the rich diversity of musical instruments and program styles that can be played back in real time, while requiring almost no overhead processing from the computer's CPU. This makes MIDI a perfect candidate for playing back soundtracks from multimedia games or from a phone (MIDI ringtone, Internet, gaming devices, etc.). As one might expect, MIDI has taken a back seat to digital audio as a serious music playback format for multimedia. Most likely, this is due to several factors, including:

- A basic misunderstanding of the medium
- The fact that producing MIDI content requires a fundamental knowledge of music
- The frequent difficulty of synchronizing digital audio to MIDI in a multimedia environment
- The fact that soundcards, phones, etc., often include poorly designed FM synthesizers (although most operating systems now include higher quality software synths)

Fortunately, a number of companies have taken up the banner of embedding MIDI within their media projects and Google's Chrome now includes integrated MIDI support within the browser itself. All of these factors have helped push MIDI a bit more into the Web mainstream. As a result, it's becoming more common for your PC to begin playing back a MIDI score on its own or perhaps in conjunction with a game or more data-intensive program.

The following information relates to MIDI as it functions within the multimedia environment. Of course, more in-depth information on the spec and its use can be found within Chapter 9, MIDI and the DAW.

Standard MIDI Files

The accepted format for transmitting music-related data and real-time MIDI information within multimedia (or between sequencers from different manufacturers) is the standard MIDI file. This file type (which is labeled with a .mid or .smf extension) is used to distribute MIDI data, song, track, time signature and tempo information to the general masses. Standard MIDI files can support both single and multichannel sequence data and can be loaded into, edited and then directly saved from almost any sequencer package. When exporting a standard MIDI file, keep in mind that they can come in two basic flavors – type 0 and type 1:

- Type 0 is used whenever all of the tracks in a sequence need to be merged into a single MIDI track. All of the notes will have a channel number attached to them (i.e., will play various instruments within a sequence); however, the data will have no definitive track assignments. This type might be the best choice when creating a MIDI sequence for a standard device or the Internet (where the sequencer or MIDI player application might not know or care about dealing with multiple tracks).

- Type 1, on the other hand, will retain its original track information structure and can be imported into another sequencer type with its basic track information and assignments left intact.

General MIDI

One of the most interesting aspects of MIDI production is the absolute uniqueness of each professional and even semi-pro project studio. In fact, no two studios will be even remotely alike (unless they've been specifically designed to be the same or there's a very unlikely coincidence). Each artist will have his or her own favorite equipment, supporting hardware, assigned patches and way of routing channels/tracks. The fact that each system setup is unique and personal has placed MIDI at odds with the need for complete compatibility in the world of multimedia. For example, if you import a MIDI file over the Net that's been created in another studio, the song will most likely attempt to play with a totally irrelevant set of sound patches (it might sound interesting, but it won't sound anything like it was originally intended). If the MIDI file is loaded into completely different setups, the sequence will again sound completely different … and so on.

To eliminate (or at least reduce) the basic differences that exist between systems, a standardized set of patch settings, known as General MIDI (GM), was created. In short, General MIDI assigns a specific instrument patch to each of the 128 available program change numbers. Since all electronic instruments that conform to the GM format must use these patch assignments, placing GM program change commands at the header of each track will automatically instruct the sequence to play with its originally intended sounds and general song settings. In this way, no matter what synth, sequencer and system setup is used to play the file back, as long as the receiving instrument conforms to the GM spec, the sequence will be heard using its intended instrumentation.

Tables 12.4 and 12.5 detail the program numbers and patch names that conform to the GM format. These patches include sounds that include synthesizer sounds, ethnic instruments and sound effects that have been derived from early Roland synth patch maps. Although the GM spec states that a synth must respond to all 16 MIDI channels, the first 9 channels are reserved for instruments, while GM restricts the percussion track to MIDI channel 10.

GRAPHICS

Graphic imaging occurs on the computer screen in the form of pixels. These are basically tiny dots that blend together to create color images in much the same way that dots are combined to give color and form to your favorite comic strip. Just as word length affects the overall amplitude range of a digital audio signal, the number of bits in a pixel's word will affect the range of colors that can be displayed

Table 12.4	GM Non-Percussion Instrument (Program Change) Patch Map	
1. Acoustic Grand Piano	44. Contrabass	87. Lead 7 (fifths)
2. Bright Acoustic Piano	45. Tremolo Strings	88. Lead 8 (bass + lead)
3. Electric Grand Piano	46. Pizzicato Strings	89. Pad 1 (new age)
4. Honky-Tonk Piano	47. Orchestral Harp	90. Pad 2 (warm)
5. Electric Piano 1	48. Timpani	91. Pad 3 (polysynth)
6. Electric Piano 2	49. String Ensemble 1	92. Pad 4 (choir)
7. Harpsichord	50. String Ensemble 2	93. Pad 5 (bowed)
8. Clavichord	51. SynthStrings 1	94. Pad 6 (metallic)
9. Celesta	52. SynthStrings 2	95. Pad 7 (halo)
10. Glockenspiel	53. Choir Aahs	96. Pad 8 (sweep)
11. Music Box	54. Voice Oohs	97. FX 1 (rain)
12. Vibraphone	55. Synth Voice	98. FX 2 (soundtrack)
13. Marimba	56. Orchestra Hit	99. FX 3 (crystal)
14. Xylophone	57. Trumpet	100. FX 4 (atmosphere)
15. Tubular Bells	58. Trombone	101. FX 5 (brightness)
16. Dulcimer	59. Tuba	102. FX 6 (goblins)
17. Drawbar Organ	60. Muted Trumpet	103. FX 7 (echoes)
18. Percussive Organ	61. French Horn	104. FX 8 (sci-fi)
19. Rock Organ	62. Brass Section	105. Sitar
20. Church Organ	63. SynthBrass 1	106. Banjo
21. Reed Organ	64. SynthBrass 2	107. Shamisen
22. Accordion	65. Soprano Sax	108. Koto
23. Harmonica	66. Alto Sax	109. Kalimba
24. Tango Accordion	67. Tenor Sax	110. Bag Pipe
25. Acoustic Guitar (nylon)	68. Baritone Sax	111. Fiddle
26. Acoustic Guitar (steel)	69. Oboe	112. Shanai
27. Electric Guitar (jazz)	70. English Horn	113. Tinkle Bell
28. Electric Guitar (clean)	71. Bassoon	114. Agogo
29. Electric Guitar (muted)	72. Clarinet	115. Steel Drums
30. Overdriven Guitar	73. Piccolo	116. Woodblock
31. Distortion Guitar	74. Flute	117. Taiko Drum
32. Guitar Harmonics	75. Recorder	118. Melodic Tom
33. Acoustic Bass	76. Pan Flute	119. Synth Drum

Table 12.4	Continued	
34. Electric Bass (finger)	77. Blown Bottle	120. Reverse Cymbal
35. Electric Bass (pick)	78. Shakuhachi	121. Guitar Fret Noise
36. Fretless Bass	79. Whistle	122. Breath Noise
37. Slap Bass 1	80. Ocarina	123. Seashore
38. Slap Bass 2	81. Lead 1 (square)	124. Bird Tweet
39. Synth Bass 1	82. Lead 2 (sawtooth)	125. Telephone Ring
40. Synth Bass 2	83. Lead 3 (calliope)	126. Helicopter
41. Violin	84. Lead 4 (chiff)	127. Applause
42. Viola	85. Lead 5 (charang)	128. Gunshot
43. Cello	86. Lead 6 (voice)	

Table 12.5	GM Percussion Instrument (Program Key Number) Patch Map (Channel 10)	
35. Acoustic Bass Drum	51. Ride Cymbal 1	67. High Agogo
36. Bass Drum 1	52. Chinese Cymbal	68. Low Agogo
37. Side Stick	53. Ride Bell	69. Cabasa
38. Acoustic Snare	54. Tambourine	70. Maracas
39. Hand Clap	55. Splash Cymbal	71. Short Whistle
40. Electric Snare	56. Cowbell	72. Long Whistle
41. Low Floor Tom	57. Crash Cymbal 2	73. Short Guiro
42. Closed Hi-Hat	58. Vibraslap	74. Long Guiro
43. High Floor Tom	59. Ride Cymbal 2	75. Claves
44. Pedal Hi-Hat	60. Hi Bongo	76. Hi Wood Block
45. Low Tom	61. Low Bongo	77. Low Wood Block
46. Open Hi-Hat	62. Mute Hi Conga	78. Mute Cuica
47. Low-Mid Tom	63. Open Hi Conga	79. Open Cuica
48. Hi-Mid Tom	64. Low Conga	80. Mute Triangle
49. Crash Cymbal 1	65. High Timbale	81. Open Triangle
50. High Tom	66. Low Timbale	

Note: in contrast to Table 12.4, the numbers in Table 12.5 represent the percussion keynote numbers on a MIDI keyboard, not program change numbers.

in a graphic image. For example, a 4-bit word only has 16 possible combinations. Thus, a 4-bit word will allow your screen to have a total of 16 possible colors; an 8-bit word will yield 256 colors; a 16-bit word will give you 65,536 colors; and a 24-bit word will yield a whopping total of 16.7 million colors! These methods of displaying graphics onto a screen can be broken down into several categories:

- *Raster graphics*: in raster graphics, each image is displayed as a series of pixels. This image type is what is used when a single graphic image is used (i.e., bitmap, JPEG, GIF, PNG or TIFF format). The sense of motion can come from raster images only by successively stepping through a number of changing images every second (much in the same way that standard video images create the sense of motion).

- *Vector graphics*: this method displays still graphic drawings using geometric shapes (lines, curves and other shapes) that are placed at specific coordinates on the screen. Being assigned coordinates, shape, thickness and fill, these shapes combine together to create an image that can be simple or complex in nature. This script form reduces a file's data size dramatically and is used with several image programs.

- *Vector animation*: much like vector graphics, vector animation can make use to the above shapes, thickness, fills, shading and computer-generated lighting to create a sense of complex motion that moves from frame to frame (often with a staggering degree of realism). Obviously, with the increased power of modern computers and supercomputers, this graphic art form has attained higher degrees of artistry or realism within modern-day film and gaming production and design.

VIDEO

With the proliferation of computers, DVD/Blu-ray players, cell phones, video interface hardware and editing software systems, desktop and laptop video has begun to play an increasingly important role in home and corporate multimedia production and content. In short, video is encoded into a data stream as a continuous series of successive frames, which are refreshed at rates that vary from 12 or fewer frames/second (fr/sec) to the standard broadcast rates of 29.97 and 30 fr/sec (or higher for 3D applications). As with graphic files, a single full-sized video frame can be made up of a gazillion pixels, which are themselves encoded as a digital word of n bits. Multiply these figures by nearly 30 frames and you'll come up with rather impressive data file size and throughput rates.

Obviously, it's more common to find such file sizes and data throughput rates on higher-end desktop systems and professional video-editing workstations; however, several options are available to help bring video down to data rates that are suitable for the Internet and multimedia:

- *Window size*: the basics of making the viewable picture smaller are simple enough: reducing the frame size will reduce the number of pixels in a video frame, thereby reducing the overall data requirements during playback.

- *Frame rate*: although standard video frame rates run at around 30 fr/sec (United States and Japan) and 25 fr/sec (Europe), these rates can be lowered to 12 fr/sec in order to reduce the encoded file size or throughput.
- *Compression*: in a manner similar to that which is used for audio, compression codecs can be applied to a video frame to reduce the amount of data that are necessary to encode the file. This is done by filtering out and smoothing over pixel areas that consume data or by encoding data that doesn't change from frame to frame into shorthand that reduces data throughput. In situations where high levels of compression are needed, it's common to accept degradations in the video's resolution in order to reduce the file size and/or data throughput to levels that are acceptable for a restrictive medium (e.g., the Web).

From all of this, it's clear that there are many options for encoding a desktop video file. When dealing with video clips, tutorials and the like, it's common for the viewing window to be medium in size and encoded at a medium to lower frame rate. This middle ground is often chosen in order to accommodate the standard data throughput that can be streamed off of most the Web. These files are commonly encoded using Microsoft's Audio-Video Interleave (AVI) format, QuickTime (a common codec developed by Apple that can be played by either a Mac or PC) or MPEG 1, 2 or 4 (codecs that vary from lower multimedia resolutions to higher ones that are used to encode DVD movies). Both the Microsoft Windows and Apple OS platforms include built-in or easily obtained applications that allow all or most of these file types to be played without additional hardware or software.

Table 12.6	Internet Connection Speeds	
Connection	**Speed (bps)**	**Description**
56k dial-up	56 Kbps (usually less)	Common modem connection
ISDN	128 Kbps (older technology)	
ISDN PRI/E1	1.5 Mbps/1.9 Mbps	
DSL	256 Kbps to 20 Mbps	
Cable	Up to 85 Mbps	
OC-1	52 Mbps	Optical fiber
OC-3	155 Mbps	Optical fiber
OC-12	622 Mbps	Optical fiber
OC-48	2.5 Gbps	Optical fiber
Ethernet	10 Mbps (older technology) up to 100 Gbps	Local-area network (LAN), not an Internet connection

MULTIMEDIA IN THE "NEED FOR SPEED" ERA

The household phrase "surfin' the Web" has become synonymous with jumping onto the Net, browsing the sites and grabbing all of those hot songs, videos, and graphics that might wash your way. With improved audio and video codecs and ever-faster data connections (Table 12.6), the ability to search on any subject, download files and stream audio or radio stations from any point in the world has definitely changed our perception of modern-day communications.

ON A FINAL NOTE

One of the most amazing things about multimedia, cyberspace and their related technologies is the fact that they're ever-changing. By the time you read this book, many changes will have occurred. Old concepts will have faded away … and new and possibly better ones will take over and then begin to take on a new life of their own. Although I've always had a fascination with crystal balls and have often had a decent sense about new trends in technology, there's simply no way to foretell the many amazing things that lie ahead in the fields of music, music technology, gaming, visual media, multimedia and especially cyberspace. As with everything techno, I encourage you to read the trades and surf the Web to keep abreast of the latest and greatest tools that have recently arrived, or are on the horizon.

CHAPTER 13

Music Notation and Printing

Over the past few decades, the field of transcribing musical scores and arrangements has been strongly affected by both the computer and MIDI technology. This process has been greatly enhanced through the use of newer generations of computer software that makes it possible for music notation data to be entered into a computer by:

- Manually placing the notes onto the screen via keyboard or mouse movements
- Direct MIDI input
- The use of sheet music scanning technology

Once entered, these notes can be edited in an on-screen environment that lets you change and configure a musical score or lead sheet using standard cut-and-paste editing techniques. In addition, most programs allow the score data to be played directly from the score using electronic instrument hard- and software via MIDI. A final and important program feature is their ability to display and quickly print out hardcopies of a score or lead sheets in a wide number of print formats and styles.

A music notation program (also known as a music printing program) lets you enter musical data into an on-screen score in a number of manual and automated ways (often with varying degrees of complexity and functional depth). Programs of this type (Figure 13.1) offer a wide range of notation symbols and font styles that can be entered either from a computer keyboard or mouse. In addition to accepting MIDI input (allowing a part to be played directly into a score), all music transcription programs will let you enter a score manually. This can be done in real time (by playing a MIDI instrument/controller or finished sequence into the program) or in step time (by entering the notes of a score one note at a time from a MIDI controller), or by entering a standard MIDI file into the program (which uses a sequenced file as the notation source).

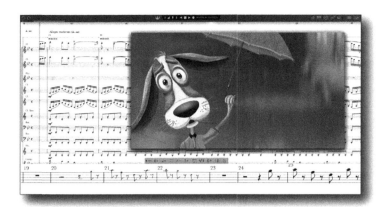

ENTERING MUSIC DATA

In addition to dedicated music printing programs, most DAW or sequencer packages will often include a basic music notation application that allows the sequenced data within a track or defined region to be displayed and edited directly within the program (Figure 13.2). A number of high-level workstations, however, also offer various levels of scoring features that allow sequenced track data to be notated and edited in a professional fashion into a fully printable music score.

As you might expect, music printing programs will often vary widely in their capabilities, ease of use and general features. These differences often center around the:

- Graphical user interface (GUI)
- Methods for inputting and editing data
- The number of instrumental parts that can be placed into a score
- The overall selection of musical symbols

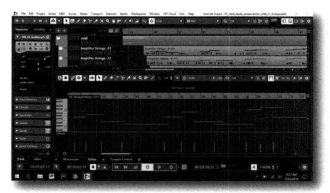

- The number of musical staves (the lines that music notes are placed onto) that can be entered into a single page or overall score
- The ability to enter text or lyrics into a score

As with most programs that deal with artistic production, the range of choices and general functionality reflect the style and viewpoints of the manufacturer, so care should be taken when choosing the professional music notation program that would work best for your needs.

SCANNING A SCORE

Another way to enter music into a score is through the use of an optical recognition program (Figure 13.3). These programs let you use a standard flatbed scanner to directly enter sheet music or a printed score into a music program, and then save the notation, data and general layout as a standardized Notation Interchange File Format (NIFF) file, which can then be exported to another scoring/notation program.

EDITING A SCORE

As was stated earlier, music notation and printing programs allow music-related data to be input in a number of ways (manually, via MIDI, file import or optical scanning). Once the data is imported into the program, it's generally a simple matter to add, delete or change individual notes, duration and markings by using a combination of computer keyboard commands, mouse movements and MIDI keyboard commands. Of course, larger blocks of music data can also be edited using the standard cut-and-paste approach. As you might expect, a wide selection of musical symbols are commonly available; these can be placed and moved within a score to denote standard note lengths, rest duration markings, accidental markings (flat, sharp and natural), dynamic markings (e.g., pp, mp, mf, ff) and a host of other important score markings that can range from the commonly used to the obscure (Figure 13.4). In addition to standard notation

FIGURE 13.3
SmartScore X2 Pro music OCR scanning/editing program. (Courtesy of Music Imaging Technologies, www.musitek.com.)

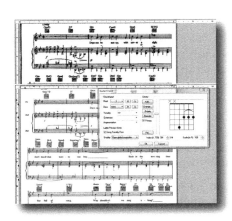
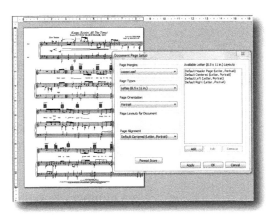

FIGURE 13.4
Dorico Advanced Music Notation System showing TrueType display of notation font symbols. (Courtesy of Steinberg Media Technologies GmbH, A Division of Yamaha Corporation; www.steinberg.net.)

entry, text can usually be placed into a lead sheet or score for the purpose of adding song lyrics, song titles, special performance markings and additional header/footer information.

One of the major drawbacks to entering a score into such a program (either as a real-time performance or as an imported file) is the fact that music notation is an interpretive art. As the saying goes, "To err is human", and this human feel is what usually gives music its full range of emotional expression. Although the algorithms are constantly getting better, it's sometimes difficult for a computer program to properly interpret these minute, yet importantly subtle, imperfections and then place them into the score in their exact and proper place. (For example, the program might interpret a section of a passage with a held 1/4 note as one that contains a dotted 1/4 or one that's tied to a 1/32 note.) In short, it will often have no way of actually knowing the note's intended placement and parameters and will have to make its best guess. As such, even though the algorithms are getting better at interpreting musical data and can make use of quantization to instruct a computer to round a note value to a specified length, a score will still often need to be edited manually by a human in order to correct for misinterpretations.

Before continuing, I'd like to point out the importance of taking time to properly set up the initial default settings of a music notation program. These can be in the form of global settings that can help set such parameters as measure widths, number of stems, instrument layouts and default key signatures. Less important decisions, such as title, composer and instrument name fonts, can also be entered beforehand. To save time in the long run, you might also consider making a few file setup templates that contain several of the more common time and key signatures, stave and instrument layouts and other settings that you tend to commonly work with.

Finally, it's often wise to take the time to familiarize yourself with the vast number of notational parameters that are contained in most programs before beginning work on a song or score. This can save you from having real head-aches when making changes (even minor setting might have an impact on the overall piece). Taking the time to lay out a composition's basic style rules before-hand can definitely help you avoid unnecessary bandage-type fixes that might be needed at a later stage in the process.

PLAYING BACK A SCORE

In the not-so-distant past, classical and even film composers commonly had to wait months or years in order to hear their finished composition. Orchestras and ensembles were generally far too expensive and often required that the com-poser have a financial patron or a good knowledge of corporate or state politics in order to get their works heard. One simple solution to this obstacle was the piano reduction, which served to condense a score down into a compromised version that could be played at the piano keyboard.

With the advent of MIDI, however, classical compositions and film scores could be played through various MIDI instruments within a composing/project studio or directly from a DAW's sequencer or notation program. In this way, a produc-tion system can approximate a working rendition of the final performance with relative ease, allowing the artist to check for errors and final tweaks before being subjected to the expense, time constraints and the inherent difficulties of work-ing with a live orchestra.

DISPLAYING AND PRINTING OUT A SCORE

Once the score has been edited into a final form, the process of creating a printed hardcopy is relatively simple. Generally, a notation program can help you to lay out the score in a way that best suits your taste, the producer's taste or the score's intended purpose. Often, a professional program will let you make final changes to a wide range of parameters such as margins, measure widths, title, copyright and other text-based information. Once completed, the final score can then be printed out using a standard, high-quality ink-jet or laser printer.

In this modern age of computer technology, it's also possible to display sheet music directly onto an iPad or dedicated LCD screen, bypassing the printing process altogether. Dedicated LCD sheet music displays (Figure 13.5) allow an entire season's worth of concerts or a semester's worth of class material to be saved within its internal memory. In addition, online sites allow for sheet music to be downloaded and taken on the road – all without the need for printing out the music. Of course, printing the music out as a backup might save the day in the case of a power blackout or the ever-present technical glitch (you know what they say about being prepared?).

In addition to the growing number of options for storing and displaying sheet music in the digital domain, several online services have begun to show up that

FIGURE 13.5
Forscore paperless sheet music reader. (Courtesy of Forscore LLC.; www.forscore.co.)

offer printed music downloads, using a model that's not unlike iTunes® and other music download services. These services often offer a free player that can be downloaded from the site and allow for purchased downloads of sheet music in a wide range of styles and at a reasonable cost. Some tunes, including classical pieces, public-domain songs and copyright-free material, are also available free of charge.

CHAPTER 14

The Art of Production

In the past, almost all commercial music was recorded and mixed by a professional recording engineer under the supervision of a producer and/or artist. With the emergence of the project studio revolution and electronic music, the vast majority of production facilities have become much more personal and cost-effective in nature. Now, with the maturation of the digital revolution, MIDI and digital audio, workspaces are now being owned by individuals, small businesses and artists who are taking the time to become more experienced in the basic guidelines of creative and commercial music and audio production practices.

Within the music production industry, it's a well-known fact that most professionals have to earn their "ears" by logging countless hours behind the console. Although there's absolutely no substitute for this experience, the production abilities and ears of electronic musicians have naturally evolved as equipment quality, media education and overall technology get better and as musicians become more knowledgeable about proper mixing environments and techniques … usually by mixing their own compositions.

In this chapter, we'll be taking a common-sense, basic look into the audio production process (Figure 14.1). The goal here is to gain a better understanding of how the technology of the music studio environment combines with the everyday practice of getting the most out of your production environment, so as to understand the underlying concepts that go into making the process as smooth and professional as possible. In short, just because your studio sits on a desk in the corner of your bedroom, doesn't mean that you can't adopt a professional approach and attitude that can lead to drastically improved results.

One of the most important insights to be gained (beyond an understanding of the technology and tools of the trade) is the fact that there are no rules for the process of recording … it's an artform. This non-rule holds true insofar as inventiveness and freshness tend to play an essential role in keeping the creative process of making music alive and exciting.

Here, I'd also like to bring a basic axiom in audio production into the equation.

"Wherever you may be, there you are".

To me, this means that it's almost always a good idea to start off at the level where you are. If you're a beginner, I'd suggest that you don't dive in and buy everything in the store. I have met special people who have done their extensive research and have gone out to buy the best equipment, in order to get started with the best possible tools. I've found, however, that this approach isn't always the best for everyone to take. Quite often, it's a better idea to start out with an affordable piece of gear that's at your particular skillset level. This'll give you the time to figure out what might be the best production choice or piece of gear for your studio when the time comes to jump to the next level. Making mistakes and taking small steps is often a big part of the process of becoming a Jedi Master.

Speaking of, another part of becoming a Jedi Master is learning to be patient and gentle with yourself at first. When I sat down at my first console, I was totally lost:

- What are all these controls?
- How am I supposed to know what sounds good?
- Should I use EQ or not and, by the way, is this compressor working?
- How is this mix supposed to sound and how do I get started?

All of this is scary stuff, and we've all been there. As with getting to Carnegie Hall and mastering anything worthwhile, it takes practice and commitment … and the former comes with time.

Oh yeah, one last thing … I'm seeing a lot of YouTube videos all over the place telling you how to "Get the perfect bass sound" or whatever. These types of instructional tools are always good and have their place; however, I'd like to add in that art is in the ears (and mind) of the beholder. Learning from others and then using those tools and techniques to develop your own style is also what the "Hokey-Pokey" is all about. Go ahead and YouTube this old song … it's rather fun!

FIGURE 14.1
Workin' it in the basement.
(Courtesy of Ableton AG,
www.ableton.com.)

THE ART OF PREPARATION

Whenever a producer is asked about the importance of preparation, more often than not they'll most likely place preparation at or near the top of the list for capturing a project's sound, feel and performance. Having a plan (both musically and technically) to get the job done in a timely fashion, with the highest degree of artistic motivation and technical preparedness will definitely help get the project off to a good start. Details like making sure that the instruments are tuned, the cables work and that the DAW is sorted out are important to take care of *before* the red light goes on. Making sure that the musicians are well-practiced, relaxed and rested doesn't hurt either … and don't forget to have water, fruit and food on hand to keep everyone at their best. You'd be surprised how the little things can add to the success of a project … preparation, baby!

At its basic level, here are five life "tools" that can best help guide you toward the completion of a successful project:

- Preparation
- Attention to detail
- A creative and open mind (being open to experimentation, while keeping to your goals)
- A common-sense approach to the overall process
- A good attitude

By far, one of the most important steps to take when approaching a project that involves a number of creative and business stages, decisions and risks is *preparation*. Without a doubt, the best way to avoid production pitfalls and to help get you, your client or your band's project off the ground is to discuss and outline the many factors and decisions that will affect the creation and outcome of that all-important "final product". Just for starters, here are a number of ever basic questions that should be asked long before the big red "REC" button is pressed:

- How much will this production cost?
- How (if at all) are you planning to recoup the production costs?
- Will others share in the costs or financial arrangements?
- Do you want to think about pre-sales and/or crowd-funding to help defray the cost?
- How is it to be distributed to the public? Self-distribution? Digital download? Indy? Record label?
- Do you need a producer or will you self-produce?
- How much practice will you need? Where and when? Shall the dates be put on the calendar?
- Should you record it in a band member's project studio or at a commercial studio?

- If a project studio is used and it works out, should you mix it at the commercial studio? Which one and who will do the mix?
- Will it be self- or professionally mastered? Who will do the mastering?
- Who's going to keep track of the time and budget? Is that the producer's job … or will he or she be strictly in charge of creative and contact decisions?
- Will you need a music lawyer to help with contacts and contracts?
- Who will be in charge of the artwork and the website?
- When should the artist or group's artistic and financial goals be discussed and put down on paper? (Of course, it's always best to discuss budget requirements, possible rewards and written contract stipulations as early as possible in the game!)

These are but a few of the issues that should be addressed before tackling a project. Of course, they'll change from project to project, might change over time and will depend upon the final project's scope and purpose; however, in the final analysis, asking the right questions (or finding someone who can help you ask the right questions) can help keep you focused, on schedule and on budget.

Now that you've answered these questions, here's a list of tasks that are often wise to tackle before going into production:

- Create a mission statement for you/your group and the project. This can help clue your audience in as to what you are trying to communicate through your art and music and can greatly benefit your marketing goals. For example, you might want to answer such questions as: who are you? What are your musical goals? How should the finished project sound? What emotions should it evoke?
- What is the budget for this project? How will it be sold? What are the marketing strategies?
- Start working on the project's artwork, packaging and website ASAP.
- You might think about copyrighting your songs. Form SR is used for the registration of published or unpublished sound recordings or a recorded performance. If you wish to register only the underlying musical composition or dramatic work, Form PA or Short Form PA should be used. Copies of and information on these and other forms can be found at www.copyright.gov/forms. Another way to protect your songs in a more automatic way is by using Creative Commons (www.creativecommons.org), which is a global nonprofit organization that enables sharing and reuse of creativity and knowledge through the provision of free legal tools for sharing and protecting the artist's works.

All of these are just but *a few* of questions that should be asked before tackling a project. Of course, they'll change from one project to the next and will depend on the project's scope and purpose. However, in the final analysis, asking the right questions (or finding someone who can help you ask them) can help keep your project and career on-track.

GETTING THE BEST OUT OF YOUR PROJECT

As I'm sure you're aware, an audio production system (as well as the process itself) is a complex chain of acoustic interactions, interconnected electrical/digital devices and artistic decisions that go together to create a final end product. Again, this chain can only be as strong as its weakest link, as stated in the "GOOD RULE" …

Good musician + good instrument + good performance + good acoustics + good mic + good placement = good sound.

Now is a good time to talk about the various acoustic and technical aspects for getting the most out of your production system. These relate to the parts of the chain that can help make or break your overall sound. Of course, there are a number of guidelines and basic steps that can be taken to get the most out of your room or production system … so let's get to them.

THE ART OF RECORDING

When being used in a book about MIDI, the term "Art of Recording" will almost certainly mean something different than if it were put into practice in a professional recording facility. The goals, methods and tools will differ, in that the instruments will primarily be electronic in nature. There will, however, be times when you'll want to lay down a vocal track, capture an acoustic guitar or maybe even record a live drum set (or smaller kit). The various situations that you might run into might not be as wide-ranging as what you'll find in an all-acoustic setting, but you should also be prepared to get the best sound out of an instrument, vocal technique or toy that might be right for your song.

The road to getting the best sound out of your production can be best traveled by considering a few simple rules:

- Rule 1: there are no rules, only guidelines. And although guidelines can help you achieve a good sound, don't hesitate to experiment in order to get a sound that best suits your needs or personal taste.
- Rule 2: the overall sound of an audio signal is no better than the weakest link in the signal path. If a mic, device or process doesn't sound as good as it could, make the changes to improve it BEFORE you commit it to disk, tape or whatever. More often than not, the concept of "fixing it later in the mix" will often put you in the unfortunate position of having to correct a situation after the fact, rather than getting the best sound and/or performance during the initial session.
- Rule 3: whenever possible, use the "Good Rule": Good musician + good instrument + good performance + good acoustics + good mike + good placement = good sound. This rule refers to the fact that a music track will only be as good as the performer, instrument, mic, mic placement and the

entire signal chain that follows it. If any of these elements falls short of its potential, the track will suffer accordingly. However, if all of these links are the best that they can be, the recording will almost always be something that you can be proud of!

Microphones

It goes without saying that having quality mics around can go a long way towards capturing vocals, instruments or almost anything that will come your way. Never before has there been such a wide range of mic types, configurations and price ranges that are available to the pro, semi-pro or home artist/producer. As I'm sure you're aware, mic types that would've cost thousands of dollars in the past can be had for hundreds or less. As with many things, it's not how much that mic (or mics) will cost that makes the difference, it's how you use it. I could go into all of the possible combinations and possible settings that you might come across, but this is beyond the scope of this book. Many of these pickup scenarios have been covered in many other books (including my own *Modern Recording Techniques*) … so, I'll leave that part up to you to read a book or to surf the Web for videos on good miking practices. Just remember, miking (as with all things audio) is an art. Over time, you'll be able to take other people's opinions and practices and then begin to work in a way that best serves you and your own projects.

The miking of vocals and instruments (both in the studio and onstage) is definitely an art form. It's often a balancing act to get the most out of the Good Rule. Sometimes you'll simply be lucky and have all of the elements; at others, you'll have to work hard to make lemonade out of a situational lemon. The best rule of all is to use common sense and to trust your own instincts.

Obviously, most project studios will want to have a few microphones around to capture vocals, instruments and maybe re-amp a MIDI instrument track to your DAW. The choice of mic type and price range is strictly up to you, as there are lots of dynamic, ribbon and condenser mics to choose from that come under the high-quality, good-bang-for-the-buck category.

The basic facts that I'd like to get across here, is that there are two schools of thought when it comes to choosing the best mics for your production studio:

1. The Allen Sides approach: Allen is an engineer, producer and owner of the Ocean Way Recording family of recording studios, but he's also well-known for his microphone collection that spans some of the most rare and coveted mics in existence. The idea behind his collection is that there will often be a specific mic that works best for a certain instrument or placement application. As we all know, this is all well and good if you're willing to spend a million dollars on your own personal set of mics. Obviously, us mere project and home studio mortals must take a different approach.

2. The general application mic collection: when it comes to suiting your basic needs, it's often best to choose several mics (Figure 14.2) that will fit your overall recording needs. For

example, you might want to start off by getting two good condenser mics of the same model (for stereo miking). These don't need to be overly expensive, but they should be chosen for their overall quality and ability to be used in a wide range of applications (such as vocals, drums, perc and the like). Next, you might want to get a dynamic and/or ribbon pair, just to round out your small, but very effective collection.

A number of books and on-line references can help you to get the best possible mic technique for the voice or instrument application that's at hand.

Although the most important guideline in recording is the above-stated "Good Rule", this doesn't mean that you shouldn't hesitate to experiment in order to get a sound that best suits your needs or personal taste. It's important to realize that the overall sound of an audio signal is no better than the weakest link in the signal path.

The Preamp

Since the output signals of most microphones are at levels far too low to drive the line-level input stage of most recording systems, a mic *preamplifier* must be used to boost its signal to acceptable line levels (often by 30 to 70 dB). In this day and age, most of the mic *pres* (pronounced "preeze") that are designed into an audio interface or production console are capable of providing a professional, high-quality sound. It's not uncommon, however, for an engineer, producer or artist to prefer a preamp which has a personal and "special sound" that can be used in a critical application to produce just the right tone for a particular application. In such a case, an outboard mic preamp might be chosen instead (Figure 14.3) for its characteristic sound, low noise or special distortion specs. These devices might make use of tube, FET and/or integrated circuit technology, and offer advanced features in addition to its common variable input gain, phantom power and high-pass filter controls.

As with most recording tools, the sound, color scheme, retro style, tube or transistor type and budget level are up to the individual, producer and/or artist … it's completely a matter of personal style and taste … and this includes the simple (and most often used) option of choosing to use the high-quality preamps that are already built into your own interface or mixer.

FIGURE 14.2

An initial, well-rounded mic (or pair of mics) can go a long way towards making quality acoustic recordings in the studio. A second pair can then help round out your choices, should the need arise. (Courtesy of Royer Labs; www.royerlabs.com.)

The Interface

Of course, a device that deserves a lot of consideration when putting together a DAW-based production system is the digital audio interface (Figure 14.4). These devices can have a single, dedicated purpose (of dealing with the conversion of audio) or they might be multifunctional in nature (offering up MIDI, audio and sync capabilities). In either case, their main purpose in the studio is to act as a connectivity bridge between the outside world of analog audio and the computer's inner world of digital audio. Audio interfaces come in all shapes, sizes and functionalities; for example, an audio interface can be:

- Built into a computer (although, more often than not, these devices are often limited in quality and functionality)
- A simple, two-I/O audio device
- Multichannel, offering eight analog I/Os and numerous I/O expansion options
- Fitted with one or more MIDI I/O ports
- One that offers digital I/O, word clock and sync options
- Fitted with a mix and/or controller surface (with or without motorized faders) that provides for hands-on mix and functional control

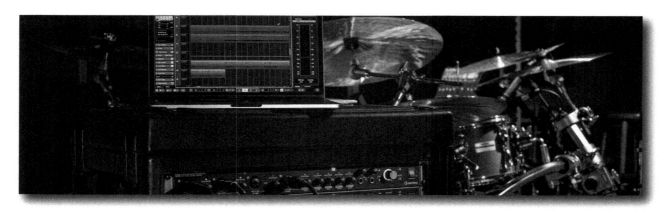

These devices might have as few as two inputs and two outputs, or it might have as many as 24 (or higher). It might offer a limited number of sample-rate and bit-depth options, or it might be capable of handling rates up to 96 kHz/24 bits or higher and can be connected to the computer via FireWire, USB or Thunderbolt.

When choosing an interface, the following considerations should be taken into account:

- The interface's overall audio quality. This should be your first consideration, as (contrary to popular opinion) no two interfaces will sound the same.
- Will it fit your studio's I/O and functional needs? Does it have expiation options that allow it to grow as your studio needs grow?

In the Production Studio

Getting back to the concept of "The Good Rule" from earlier in this chapter, one of the main goals of the recording process is to capture the best performance, using the best tools that you can. Here are a few guidelines:

- When recording acoustic instruments, it's always a good idea to assess the current situation and then go about using the tools, toys and techniques to capture the performance in the best way possible. This will involve using your best judgment for room placement, mic placement, mic choice and the like.
- Try to be aware of the musician's needs. Do they have any mic choice and/ or placement preferences? Is there water and/or snacks? Sometimes it's the little things, a good attitude and general awareness that can go a long way towards capturing the best performance from an artist.
- On the technical side, always be aware of your overall signal gain levels. By this, I mean making sure that your mic preamp levels are adequate, but well under clipping and general distortion levels. This optimum level strategy should then be followed throughout the entire signal chain, ensuring a high-quality, low-noise, low-distortion audio path (Figures 14.5 and 14.6). Speaking of levels, in this day and age of digital DAWs, it's definitely a good idea to keep your recorded track levels well below the distortion limit (–7 or –12 VU is often a good maximum peak level). Most self- and

FIGURE 14.5
Of course, signals that are too low at any part of the signal chain *might* add noise. Signals that are too high anywhere in the chain have a good chance of adding distortion, and signals that fall in the optimal operating range give you the best chance of getting a good, undistorted signal.

too low

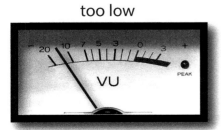

too high

just right

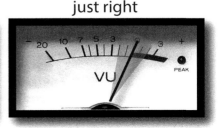

FIGURE 14.6

It's extremely important to be aware of your overall gain levels throughout the recording and DAW signal chain flowing from the mic – to the preamp/interface – through the mixing stage – through the DSP and mastering phase (if it's integrated into the project) – and finally through to the main mix out.

system-noise of a DAW recording at 24/48 or higher will be so low as to eliminate the need for recording at "hot" signal levels. Literally, there's no benefit to this in a digital system and the possibility of clipping distortion will only increase.

Documentation

There are few things more frustrating than going back to an archived session and finding that no information exists as to what instrument patch, mic type or outboard effect was used on a session (even one that's DAW-based). The importance of documenting a session within the named tracks, separate written document or within your DAW's notepad apps simply can't be overemphasized. The basic documentation that relates to the who, what, where and when of a recording, mixdown, mastering and duplication session should include such information as:

- Important session calendar dates and locations: notating special session dates, booking rates and billing info can be helpful towards keeping you on schedule and on budget. Additionally, keeping track of locations and studio website information can be helpful when putting your notes into the project's webpage.
- Artists, engineers and support staff who were involved with the project: within a session document, you might include the names, addresses and other pertinent info about the staff and musicians who were involved with the project. Their names and duties can then be placed into the liner and/or Web notes for this project. This (and other session notes) can then be saved into a (docs) folder within the session's song or overall directory.
- Session tempo: although the tempo will probably be programmed into the DAWS's project file, you will probably want to notate the original tempo. Of course, you'll want to notate any changes in tempo that might take place over the course of the overall project's life.
- Session calendar dates and locations.
- Original and edited song tempos, signatures, etc.
- Mic choice and studio placement: taking pictures with your phone or camera can be a big help towards reconstructing parts of a session or overdub, should you have to go back and make any changes. These pics should then

be placed within a (pics) directory within the song or overall session directory. In addition, you could also draw up a studio diagram to help you remember artist, equipment and mic placement within the room.

- Outboard equipment types and their settings: if external hardware instruments or effects devices are used within a production, it's almost always wise to document the name of the device and the various settings that are used within the production. Here, it's usually a good idea to get out your trusty phone or camera and snap pictures of any front-face settings that you might want to remember for later recall.

- Plugin effects and their settings or general descriptions: for the most part, all plugin setting will be saved within the DAW session that you're working on. However, there are definitely times when Murphy's Law comes into play and the plugin was inadvertently deleted or when settings on a new version might not fully support your older plugin. In short, you can't always predict how software plugin will react at a future time … so, at these times, I would recommend that you make a screenshot of the plugin and save the shot within the pics directory.

To ease this process, you might again pull out a camera or camera-phone and start snapping pictures to document the event for prosperity, as well as for technical reasons. It's often a wise idea, for social media reasons, to use your camera or bring along a high-quality video camera to quietly document the setup or actual session for your fans as extra "behind-the-scenes" online video content.

The more information that can be archived with a session (and its backups), the better the chance that you'll be able to duplicate the session in greater detail at some point in the future. Just remember that it's up to us to save and document the music of today for the fans and future playback/mix technologies of tomorrow. Basic session documentation guidelines can be found in the Guidelines & Recommendation section at (search the Web on Grammy/producers-and-engineers).

NAME THAT TRACK

When recording an audio or MIDI track to a DAW, it's always wise to name the track before the Record button is ever pressed. This simple habit lets you do the naming, instead of having the workstation assign an arbitrary track title. For example, giving the track a name of "johnsbass" will probably cause the DAW to name the recorded track as johnsbass 01 and subsequent takes as johnsbass 02, johnsbass 03, etc. If you don't name it at the outset the DAW might give it a name like "track 12001." Obviously, finding the alternative take that's named "johnsjazzbass9" would be easier to find in a crowded audio directory than "track 12009". You might want to keep these names as short as possible (or keep the first few alphanumerics as descriptive as possible), as the track display on most workstations and controller readouts will allow only eight or so characters per name and will drop the remaining ones.

ORGANIZE YOUR COLORS

Almost every workstation will let you assign a color to a track or grouping of tracks from a selectable color palate (Figure 14.7). This handy feature makes it much easier to visually identify and organize your tracks according to instrument groups or types. Should you want, you could create and save a document that lists your favorite instrument groups and assign a color code to each, so as to have a consistent color scheme.

VISUAL GROUPINGS AND SUBFOLDERS

Certain DAWs will allow any number of tracks and/or media track types to be organized into physical track folders (Figure 14.8) that can be collapsed or expanded at will. In this way, tracks can be dragged into a folder that relates to its specific category (e.g., drums, background vocals, strings). The obvious advantage to this is the ability to quickly and visually identify the groups. Most DAWs allow the tracks within the folder to be either expanded or collapsed. Of course, whenever a folder is expanded, all of the tracks will be displayed in the edit window. When collapsed, we might only see the named folder track (which might or might not contain colors that represent the folders that are within it).

Collapsed folders can actually be a very useful tool towards helping to keep the session looking organized. For example, we might have a number of MIDI sequence tracks that are ready to export to their respective audio tracks. Once exported, the MIDI tracks can be deactivated (or simply not routed to an instrument) in the menu and then moved to the MIDI folder within the session. In this way all of the visual clutter of these important and saved tracks can be minimized into a single collapsed folder. The same can be said for software instrument plugins … Once you've exported the tracks to audio (if you prefer to work in this way), you can then deactivate them (thereby keeping the settings, etc., within memory, but eliminating any unnecessary load on the CPU) and then move the instrument tracks to an "INST" folder. If you ever want to view or reactivate any MIDI or INST tracks, simply expand the folder to display all of the individual tracks and you're back in biz – it's that easy!

FIGURE 14.7
Color (with the palette shown highlighted) can be a helpful organization tool within a DAW project. (Courtesy of Steinberg Media Technologies GmbH, A Division of Yamaha Corporation; www.steinberg.net.)

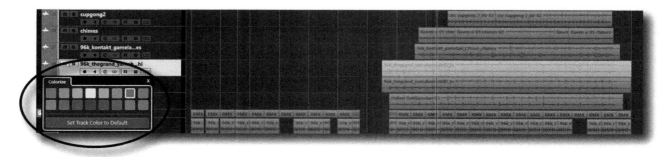

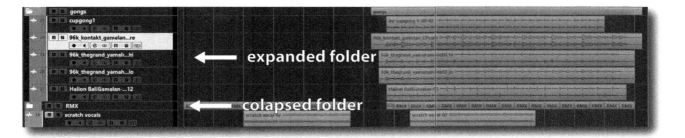

THE ART OF MIXING

Actually, the topic of the art of mixing could easily fill a book (and there are a number of good ones out there); however, I'd simply like to point out the fact that mixing is indeed an art form – and, as such, is a *very* personal process. I remember the first time that I sat down at a console (an older Neve 1604). I was truly petrified and at a loss as to how to approach the making of a mix. Am I over-equalizing? Does it sound right? Will I ever get used to this sea of knobs? Well, folks, as with all good things, the answers come with practice and dedication. It's a foregone conclusion that over time you'll begin to develop and master your own art and style of mixing.

In short, mixing is:

- First and foremost, the art of active listening.
- The art of making decisions based upon what you hear, and then acting upon these decisions.
- The process of blending art and a knowledge of audio technology and signal flow so as to turn these decisions into a technological reality, thereby creating an artistic vision.
- An art … as such, it's very subjective and individual. There is no right or wrong way and no two people will mix in exactly the same way. The object of the process is to create a mix that "frames" (presents and/or shows off) the music in the best possible light.

Ear Training

Of course, the art of listening requires that you take the time to actually listen. One of the things that a mix engineer will be doing throughout his or her career is actively listening to a mix, often over and over. This allows us to become familiar with the nuances of the song or project. Often, the instant recall aspect of a DAW gives us the ability to keep going back to a mix and improve it ad-infinitum. This is often especially true of artists who mix their own work (the "it's never done until it's perfect" category). Trust me, I'm a pro at going back and revisiting mixes, so to this, I say: go easy on yourself. Music is a process of self-discovery and expression.

FIGURE 14.8
This figure shows an expanded folder track (showing the included tracks) and the collapsed folder tracks (with the tracks hidden within the folder). (Courtesy of Steinberg Media Technologies GmbH, A Division of Yamaha Corporation; www.steinberg. net.)

In the 21st century, one of the things that we've gotten really good at is having media readily at hand. Quite often, these media are omnipresent and totally disposable. You walk into a store … there's music! A car passes you by … there's loud music! You go practically anywhere … well, you know the deal all too well. My point is that we're used to being passive in the process. Actual mixing requires that we become "active" listeners. As such, it's often a good idea to sit down with your favorite desert island songs, albums, LPs … and play them over your best speaker system. How does it sound when you actively listen to them? Then, you might take the time to listen to them over your favorite headphones. How does this change the experience for you? What can you learn from the music and their mixes?

In the end, just as with learning an instrument or doing anything well, the fact remains that as you mix, mix and mix again … you *will* get better at both your listening and mixing skills. It's a matter of experience combined with the desire to do your best.

Mixing and Balancing Basics

Once all of the tracks of a project have been recorded, assembled and edited, it's time to put the above technology to use to mix the tracks of your project into their final media forms. The goal of this process is to combine audio, MIDI and effects tracks into a pleasing form that makes use of such traditional sound tools as:

- Relative level
- Spatial positioning (the physical panned placement of a sound within a stereo or surround field)
- Equalization (affecting the relative frequency balance of a track)
- Dynamics processing (altering the dynamic range of a track, group or output bus to optimize levels or to alter the dynamics of a track so it fits within a mix)
- Effects processing (adding reverb-, delay- or pitch-related effects to a mix in order to augment or alter the piece in a way that is natural, unnatural or just plain interesting)

In addition, sounds can be built up and placed into a sonic stage through the use of natural, psycho-acoustic and processed signal cues to create a pleasing, interesting and balanced soundscape. It's obviously evident that volume can be used to move sound forward and backward within the sound field and that relative channel levels can be used to position a sound within that field. It's less obvious that changes in timbre (often but not always introduced through EQ), delay and reverb can be used to move sounds within the stereo or surround soundscape. All of this sounds simple enough; however, the dedication that's required to hone your skills within this evolving art is what mixing careers are truly made of.

PREPARATION

Just as preparation is one of the best ways to ensure that a recording session goes well … the idea of preparing for a mix can help make the process go more smoothly and be more enjoyable.

From a mix standpoint, preparing for this all-important stage definitely begins in the recording phase. For example, the phrase "fix it in the mix" stems from the 1980s, when multitrack recording started to hit its full stride. It refers to the idea that if there's a mistake or something that's not quite right in the recording … "don't worry about it; we don't have to make that decision right now. We'll just fix it later in the mix". Although to some degree, this might (or might not) be true, the fact is that this mentality can lead to gaps or just plain sloppiness in the final recording. The real problems, however, happen when multiple "fixes" aren't dealt with that begin to add up over time. If this happens, the mix can take on a not-so-great life as something that needs to be wrestled to the ground, in order to sound right. In my experience, this attention to the details at the outset during the recording phase is what separates the pros from the rest of the pack.

Although each project has a life of its own, here are just a few of the ways to prepare a mix:

- Strive to capture the artist and performance to disk or tape in a way that best reflects everyone's artistic intentions. Indeed, decisions such as EQ, EFX, etc., can be made during the mix … but if the life, spirit and essence of the musical expression isn't captured, often no amount of processing during the mix will help.
- Whenever possible, deal with the musical or technical problem as it occurs … during the recording or production phase.
- During a punch in or comp session (combining multiple takes into a single, best take), take care to match the tracks as carefully as possible. This might involve documenting the mic and instrument that was used, its placement and distance in the room to the artist, as well as any other detail that will ensure that the tracks properly blend.
- Create a rough mix during the recording phase that will often help you get started towards the final mix. This is especially easy (and almost unavoidable) to do in this age of the DAW, as the mix can begin to take shape during production in a way that can be saved and easily recalled within the session file.

There are no rules for approaching a mix; however, there are definitely guidelines. For example, when listening to a mix of a song, it's often best to listen to its overall blend, texture and "feel". A common mistake amongst those who are just beginning their journey into mixing would be to take each instrument in isolation, EQ it and try to sculpt its sound while listening to that track alone. When this is done, it's quite possible to make each instrument sound absolutely perfect on its own, but when combined into the overall mix, the blend will quite likely not work at all. This is because of the subtle interactions that occur when all of the elements are combined. Thus, it's often a good idea to first listen to the tracks within the context of the full song … and then, you can go about making any mix changes that might best serve it.

REMEMBER: the mix always has to support the song … It should bring an energy to the performance that allows the musical strengths and statements to shine through.

Actually, I lied when I said that there are no rules to the art of mixing. There is one big one – watching your overall gain structure within the session. This was discussed a bit throughout this book, but it's worth forewarning you about the perils of setting your record and/or mix faders at levels that could cause distortion. It might be obvious, but the reality of adding just a little bit more here, and a little bit more there will all begin to add up in a mix. Before you know it, your track channels, a hidden plugin, sub groups and your main outs will start to redline. Simply being aware of this natural tendency is your best defense against a distorted mix.

Preparing for a mix can come in many forms, each of which can save a great deal of setup time and frustration and help with the overall outcome of the project. Here are but a few things that might be discussed beforehand:

- Is the project ready to be mixed? Has sufficient time and emotional energy been put into the vocals? Quite often, the vocals are recorded last … leaving one of the most important elements to the last minute, when there might be time and budget restraints. This leads us to recording rule #2 … "Always budget enough time to do the vocals right, without stress".
- Will you be mixing your own project or will someone else be mixing? If it's your own project and it's in-house, then you're probably technically prepared. If the mix will take place elsewhere, then further thought might be put into the overall process. For example, will the other studio happen to have the outboard gear that you might need? Do they have the plugins that you're used to or need for the session? If not, then it's your job to make sure that you bring the installs and authorizations directly to get the session up and running smoothly … or print the effects to another track.
- If you'll be mixing for another artist and/or producer, it's often helpful to fully discuss the project with them beforehand. Is there a particular sonic style that they're after? Should the mix be aggressive or smooth sounding? Is there a particular approach to effects that should be taken?

OK … let's take a moment to walk through a fictitious mix. Remember, there's no right or wrong way to mix as long as you watch your levels along the signal path. There's no doubt that, over time, you'll develop your own sense of style. The important thing is to keep your ears open, care about the process and make it sound as good as you can.

- Let's begin building the mix by setting the output volumes on each of the instruments to a level that's acceptable to your main mixer or console. From a practical standpoint, you might want to set your tracks to unity gain or to some easily identifiable marking.
- The next step would be to begin playing the project and change the fader levels for any instrument or voice until they begin to blend in the mix. Once done, you can play the tracks to see how the overall mix levels hold up over the course of the song.
- Should the mix need to be changed at any point from its initial settings, you *might* turn the automation on for that track and begin building up the mix. You might want to save your mix at various stages of its development under a different name (MySongMix 001, MySongMix 002, etc.). This makes it easier to return to a point where you began to take a different path.
- You might want to group (or link) various instrument sections, so that overall levels can be automated. For example, during the bridge of a song you might want to reduce the volume on several tracks by simply grabbing a single fader … instead of moving each track individually.
- This calls to mind a very important aspect of most types of production (actually it's the basic tenant of life): keep it simple! If there's a trick you can use to make your project go more smoothly, use it. For example, most musicians interact with their equipment in a systematic way. To keep life simple, you might want to explore the possibility of creating a basic mixing template file that has all of the instruments and general assignments already programmed into it.
- Once you've begun to build your mix, you might want to create a rough mix that can be burned to CD or USB stick. Take it out to the car, take it to a friend's studio, put it in your best/worst boom box, have a friend critique it. Take notes and then revisit the mix after a day. You'll be surprised at what you might find out.

HUMAN FACTORS

Given the fact that mixing a song or project is an artform, by its very nature it is a very subjective process. This means that your outlook, health, mood and the very way that you perceive a song will affect your workflow as you approach the mix. Therefore, it's often a good idea to take care of yourselves and your bodies throughout the process.

- Try to be prepared and rested as you start the mix process. Working yourself too hard during the recording phase and then jumping right into the mix just might not be the best approach at that point in time.
- By the same token, over-mixing a song by slaving behind the board for hours and hours (or days and days) on end can definitely affect the way that you perceive a mix. If you've gone all blurry-eyed and your ears are tired (the "I just can't hear anything anymore syndrome") … obviously, the mix could easily suffer. This is definitely a time to point out that you might consider saving your mixes under different version names (i.e., mymix_023), so that you could go back to a previous version, should problems arise. Above all, be patient with yourself.

- You might want to take breaks … sometimes looooooongg ones. If you're not under any major time constraint, you might even consider coming back to the mix a day or even a week later. This can give us a fresh perspective, without ear fatigue or any short-term thoughts that might cloud our perception. If this is an option, you might try it out and see if it helps.

A dear friend within the Grammy organization once said to me: Dave, the one thing that I've found amongst all of the really good engineers, is the fact that they are seeking is that "perfect sound" … it's something that they "hear" in their heads, but are never quite able to reach that Holy Grail. From a personal standpoint, I can say that this is true. I'm always sonically reaching for that killer sound, that's often just beyond reach.

This again brings me back to: "Wherever you may be, there you are!" By this, I mean: we all have to start somewhere. If you're just starting out, your level of mix sophistication will hopefully be different than after you've had several years of intensive mixing experience under your belt. Be patient with yourself and your abilities, while always striving to better yourself … always a fine line to walk.

GETTING TOO CLOSE TO THE MIX

With all of the tools and techniques that are available to us, combined with the time that it takes to see a good recording and mix through to the end, there's one danger that lurks in the minds of all producers and musicians (even more so, if it's your own project) … the fact that we might simply be too close to the mix. I speak from a high degree of experience when I say that by the time I near the finish of a project and have made so many careful adjustments to get that right production and sound, there are definitely times when I have to throw my hands up and say "Is it better or not? Truthfully, there are times that I get so close to 'my' mix and 'my' music that I simply can't tell!" When this happens to you, try not to fret (too much) … it's all part of being in the production club. You might simply need a break, go for a walk or just take some time (maybe a long time) to get away from it … or you might have to just plow through to get it done on time. Just realize that this feeling of being slightly lost in your own "feelings and self-doubt" is totally normal. Here are some thoughts on this:

- Listen to your mixes on another speaker system or another system in another room altogether.
- Leave the room and listen to your mix from outside the open studio door. Many people (including myself) swear by this additional monitoring option.
- Bring a friend in to listen and give his or her thoughts.
- Take a break (this could be an hour or a week) before diving back in. If you don't have to punch the clock, time can be your friend.

I would like to close this section with another important story. Years ago, I was presenting one of my projects at an event in LA in a room that had killer speakers but really bad acoustics. All the other presenters were multi-Grammy winners; I was the only one who simply had nominations. Towards the end, I went up to the sponsor's wife and told her about my disappointment about how I thought my project sounded, compared to the others. She promptly turned to me and said – "Dave, everyone else came up to me and said the exact same thing. That's why you're here. You're always striving to make your work sound better!" That really struck a nerve. By this, she meant, if I were cocky and said "Dude, I'm the best, I totally rocked and it's perfect" (especially if it's not), I might get respect at the lower ranks, but not from the pros who know better. Moral of the story – do your best and *always* strive to do better, but to this I would add … don't beat yourself up along the way; you're only human. Perception is in the ear of the beholder and getting a mix or master right isn't always easy … especially if it's *your* art that's at stake.

THE ART OF MASTERING

Using its original definition, the mastering process is an art form that uses specialized, high-quality audio gear in conjunction with one or more sets of critical ears to help the artist, producer and/or record label attain a particular sound and feel before the recording is made into a finished manufactured product. A mastering engineer or experienced user's job is to go about the task of shaping and arranging the various cuts of a project into a final form that can be replicated into a final salable product, using three basic stages:

- Level and general balance management
- Song sequencing
- Authoring

Of course, in recent times, the general definitions and tasks of this art have changed. With the recent arrival of do-it-yourself mastering software and with an ever-expanding educational awareness of the subject, many producers and artists have begun to take of this important stage in the production process themselves.

On the subject of DIY mastering, I'm not going to offer an opinion. In fact, I've taken the DIY route myself for at least a decade now. My personal reason for this is that I'm a control freak … I simply haven't found a mastering engineer that trust enough. I've tried out several and wasn't happy with the results … so I began rolling my own.

Before we start delving under the hood of the mastering process itself, I do want to offer up a warning, or at least a thought that you should consider before you take on the self-mastering role. This refers back to the "Wherever you may be, there you are" adage. Quite simply, are you experienced enough or dedicated enough to master your own projects?

Before you give a quick answer, I'll now offer up one big opinion that I have on this subject. I believe that mastering (especially modern-day mastering) is a true artform, in a way that's equal to the recording and mix phases within a project.

As such, I believe that it's something that shouldn't be taken lightly at all. In fact, when I started my own quest into self-mastering, I actually made an agreement with myself to take at least two years to learn the craft. I feel like I've succeeded in the task, as it's gotten me four Grammy nominations; however, I'm the first to reject the idea that I am an actual mastering engineer. I've simply gotten good at mastering my own work. Hanging a mastering shingle out on the door would be an entirely different matter.

Level and General Balance Management

Of course, the final mix is where all of the magic of balancing levels, panning and effects of a song or audio program takes shape … and this is true for each of the songs within an album. Not all mix engineers are always fully aware of what's needed regarding the presence, dynamics and general "feel" in order to present the final mass-market product in its best light. This is where the first stage of the mastering process comes into play, with the finessing of the:

- General presence or lack thereof – this is done through the use of equalizers and other frequency-sculpting tools.
- Dynamics – through careful and judicious use of such dynamic range changers as compressors and limiters.
- General "feel" – that magical element called intuition and experience, which works in conjunction with a surgically designed audio system that allows the mastering engineer to use his or her sonic tools to sculpt the program material into its final form.

The subject of how high to set the levels (or more accurately, relative dynamic levels using compression/limiting) is a subject that can (and has) been the subject of many a heated debate. The major question here is – "Can it ever be too loud?" In actuality, the question that we're really asking is, "How much dynamics can we take out of the music before it begins to lose its sense of life?"

Traditionally, the industry as a whole tends to set the average level of a project at the highest possible value. This is often due to the fact that record companies will always want their music to "stand out" above the rest when played on the TV, radio, mobile phone or on the Web. This is usually accomplished by applying compression to the track or overall project. Again, this is an artistic technique that often requires experience to handle appropriately.

I'm going to stay clear of opinions here, as each project will have its own special needs and "desires", but several years ago, the industry became aware of a top-selling platinum album that was released with virtually no dynamic range at all. All of the peaks and dynamics were lost and it was pretty much a "flat-wall" of sound. This brought to light the argument that maybe, just maybe, there might be limits to how loud (how compressed) a project should be. Of course, everyone wants to be heard in the playlist in an elevator, on the phone, etc., but in reality, some degree of dynamics control is necessary to keep your mix sounding present and punchy. A good mastering engineer, however, can help you be aware that over-compression can lead to audible artifacts that can "squash" the life out of your hard-earned sound. In fact, light to moderate compression or limiting might also be the best alternative for certain types of music. Alternatively, classical music lovers often spend big bucks to hear a project in its full dynamic glory. In the end, it all depends on the content, the context and the message.

Song Sequencing: The Natural Order of Things

Once the mixdown phase has been finished, the running order in which the songs of a project are to be played, as well as the timing lengths (silence) between the tracks, will often affect the overall flow and feel of a project. The considerations for song order might be thought of as part of the mastering process that's best done by the artist, producer (those who are closest to the project) and/or the mastering engineer. This part is infinitely varied, and can only be garnered from experience and/or having an artistic "feel" for how their order and timing interactions will affect the final listening experience. A number of variables that can directly affect the sequenced order of a project include:

- Running order: which song should start? Which should close? What order feels best and supports the overall mood and intention of the project?
- Transitions: altering the transition times between songs can actually make the difference between an awkward silence that jostles the mood and a transition that upholds the pace and feel of the project. In this online era of streaming songs, this might not be so great of a consideration. However, for those of us who assemble songs into a finished album experience, it can be far more important. With regards to the Red Book CD, this standard calls for 2 seconds of silence as a default setting between tracks. Although this is necessary before the beginning of the first track, after the first track, any amount of silence can be inserted between subsequent tracks. Most editing programs will let you alter the index space timings between tracks from 00 seconds (butt splice) to a longer gap that can help to maintain the appropriate mood.
- Cross-fades: in certain situations, the transition from one song to the next is best served by cross-fading from one track directly to the next. Such a fade could seamlessly intertwine the two pieces, providing the glue that can help convey any number of emotional ties.
- Total length: how many songs will be included on the disc or album? If you've recorded extra songs, should they be included on the album, or be added as extra bonus tracks? Is it worth adding a weaker song, just to add additional time? Will you need to drop songs for the vinyl release? Again, this is generally not an issue for an online digital release, but might easily be an issue for physical release.

Authoring

Once all of the presence, dynamics, timings and general decisions have been made and taken care of, the final task at hand is to author the media into its final form. For use with downloadable music sites, this process might be as simple as providing the distribution service with as high a quality as you can. For example, if the session was done at hi-def rates (24/96 Wav, for example) and the service can accept these rates, it's best to export and upload your song or album at its native hi-definition rate. If it was recorded/mixed at 24/48 or 24/96, then you would export and upload at that rate. If the service requires the use of special coding or certain standards, it's often wise to research the matter more fully or to ask for help from a professional.

METADATA

There is one more stage that is part of the authoring process that should never be overlooked: the need for encoding accurate *metadata* into the sound file(s). Metadata (Figure 14.9) is the descriptive information that should be encoded within the sound file, giving information about the song title, artist, composer, publisher, copyright, genre, release year, song number, lyrics and any other additional comments.

It's important to pay CAREFUL attention to the details and accuracy of your metadata, as this will most likely be the descriptive text by which the song or other media can be found on the Web, over a streaming or other media service provider, as well as your own phone or computer-based media player.

Without accurate metadata, it's safe to say that the chances of anyone finding your song on the Web or media player become literally like finding a needle in a haystack … not good for you or your fans.

The Two-Step (Integrated) Mastering Option

FIGURE 14.9
Embedded metadata file tags can be added to a media file via various media libraries, rippers or editors and then viewed by a media player or file manager.

One of the concepts of traditional mastering that has always felt foreign to me lay in the fact that traditional mastering is (by its very nature) a two-step process: mixing and then mastering. The final mix is completed by the artist/engineer/producer and then the masters are sent off to another person to be finalized in a separate stage. I know that the idea is to have a professional who is emotionally detached take your "baby" and make it even better. Personally, I've never been comfortable with this traditional approach. I know this flies in the face of the

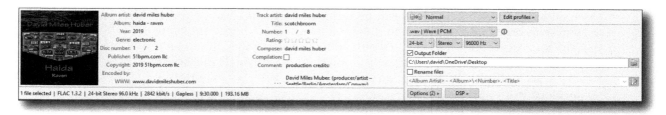

traditional mastering concept, but I am far more comfortable with a single-step, self-mastering process. This is largely due to the fact that I'm a control freak (people usually say this like it's a bad thing). To me, I feel that including the final mastering process into the DAW's song structure makes much more sense. Here's why:

- It gives me complete and total control over the entire process, with total recall to go back and to change something in the mix or mastering of a song if I want to at any time. This can be a very good thing or it can be a never-ending pitfall, if you're not careful.

- It lets me include the pre- and post-song timings within the session itself. This is done by globally moving all of the sound files in the session to the right beginning time and/or by moving your "out" export marker to a time that gives you the right amount of silence between the tracks. (Note: for certain media codecs, it's often a good idea to add a second or two of silence at the beginning, so that the player can have time to properly play the beginning beat.) Once these timings have been done you can just drop the song (or whatever) into the project cue and that's it.

- Most importantly, the final mastering chain will be included within each song, allowing any appropriate amount of final master EQ, compression and processing to be added to each song.

In short, all of the above options allow us to have complete control and repeatability (recall) of not only the recording and mix, but also the final mastering phase. It can really help should quick changes be needed (something that's really hard to do with the two-step process).

However, I'm also a strong believer that this is not for everybody. If you feel you're not qualified, or don't have the time/interest/equipment to make the journey … then hiring a pro is your best bet. Like I said, it's taken me over two years to feel that I'm qualified to master my own music, and I'd never feel qualified to master other people's mixes … but that's just my personal decision.

THE ART OF BACKING UP

The phrase "nothing lasts forever" is especially true in the digital domain of lost 1s and 0s, damaged media, dead hard drives and lost data … you know, the "Oh @#$%! factor". It's a basic fact of life that you never quite know what lies around the techno corner and, of course, it's extremely important that you protect yourself as much as is humanly possible against the inevitable screw-up. This type of headache can of course be partially or completely averted by backing up your active data, program and media files.

As a general beginning point, it's almost always wise to keep your computer's operating system and program data on a separate hard drive (usually the boot drive) and then store your data, media and session files on a separate drive. Beyond this premise, the basic rules of hard-disk management are extremely personal and will often differ from one computer user to the next (Figure 14.10). Given these differences, I'd still like to offer up some basic guidelines:

SSD

- Operating System
- Programs
    - Music Programs
    - General Programs
- My Data
    - Documents
    - Graphic Files
    - MP3 Music
- Program Archives
- Driver Archives

HDD (main drive)

- Music Project #1
    - Session #1
    - Session #2
        - Song #1
        - Song #2
        - Song #3
- Music Project #2
    - Session #1
        - Song #1
        - Song #2

HDD (backup X2)

- Music Project #1
    - Session #1
    - Session #2
        - Song #1
        - Song #2
        - Song #3
- Music Project #2
    - Session #1
        - Song #1
        - Song #2

FIGURE 14.10

Data and hard-drive management (along with a good backup scheme) are extremely important parts of media production house cleaning.

- It's important to keep your data (of all types) well-organized, using a system that's both logical and easy to follow. For example, program install files, hardware drivers and the like might be placed into their own directories; personal photos, music and other media should be kept in their own, separate place, while sound files and other session-related directories and subdirectories relating to your studio can be placed in their own special place. You get the picture.

- For those sessions that contain MIDI tracks, it's ALWAYS wise to keep these tracks within the session (don't delete them). These tracks might come in very handy during a remix or future mixdown. Moving these tracks to a folder named "MIDI" and then collapsing that folder can help reduce session clutter.

- Both Windows and the Mac OS will often try to save various media and other types of datafiles into pre-determined directories. This might work for most general users, but for more advanced users, you might *consider* placing your precious data into directories that make sense to you and that can be easily and quickly found.

- Session data should likewise be logical and easy to find. Each project should reside in its own directory and each song should likewise reside in its own subdirectory of that session project directory.

- Remember to save various take versions of a mix. If you just added the vocals to a song, go ahead and save the session under a new version name. This acts as an "undo" function that lets you go back to a specific point in a session. The same goes for mixdown versions. If someone likes a particular mix version or effect, go ahead and save the mix under a new name or version number (my greatest song 1 ver15.ses) or (my greatest song 1 ver15 favorite effect.ses). In fact, it's generally wise to save the various versions throughout the course of the mix. These session files are usually small and might save your butt at a later point in time. As a suggestion, you might want to create a "mix back" subdirectory in the session/song folder and move the older session files there, so you don't end up being confused with having 80 backup take names in your session directory.

With regards to actual backup strategies, a number of options exist. In this day and age, hard drives are still the most robust and cost-effective way of backing up your precious data. Here are some options for getting the job done, although you may have better ones that work for your own application and working scale:

- Main data storage drive: in addition to your OS drive, you should consider a second, large-capacity drive that is dedicated to your various data and media files. This will be your data Fort Knox that is divided into referenced directories that can be individually backed up. For example, it might be divided into sound files, music files, personal media, docs and the like. In this way each directory can be individually backed up to another drive, when needed.
- Secondary backup clone drive: this drive should be a copy of your main storage drive. It could also reside in your tower (if you are using one); however, I've found that a single USB portable hard drive (a 5-TB drive, in my case) also works wonders.
- Third, off-site backup clone drive: it has often been said that you're not fully backed-up unless you have a third copy that doesn't reside in your house or studio. In my case, I have another 5-TB portable hard drive that's a complete copy of the main and secondary drives that's placed off site in another safe place. This could be stored in your home (if you work away from home), your parent's house, a friend's house or in a secure bank vault box.
- Cloud storage: of course, this third off-site clone could be saved to the cloud or other storage network. The downfall here would be the time that's required to up- and download the data. On the positive side, you'd be able to access the data from literally anywhere and the data could be shared over multiple computer systems.

Seriously, apart from my family, the most important and non-replaceable thing that I own resides on those drives … and I don't treat this backup responsibility lightly. The frequency with which one should back up depends upon your backup scheme and how important the data is (actually, they're all important). Personally, I'd make sure that the entire system matches up perfectly every three months … less if possible.

THE ART OF DISTRIBUTION, MARKETING AND SALES

We've all known that the business of music has changed. Some think for the better, some for the worse. It all depends on your perspective. In the past, if you were a bedroom producer from the Corn Belt, you wouldn't have much of a chance in hell of getting your music out to the public, as you'd have to go through the process of finding an established record label that would be willing to take a chance on you. Now, as we all know, it's as easy as uploading your mix to an on-line distributor and BOOM, the masses can now find you on all of the major streaming services.

The problem with this new level playing field is that it's largely level. Since everyone is able to find and upload music on the Web ... everyone does! This means that it's harder than ever to rise above the noise to get your stuff heard, when you're competing against millions of other artists' downloads and playlists.

Obviously, the old adage of "If you release it, they'll buy it" doesn't work. The downside of the new digital age of self-distribution is that you have to do self-marketing as well. It's always been the case that once the music has been sculpted into its final, marketable form ... only then does the real work begin. All of us who have released our own productions have (hopefully) learned this lesson. I am learning, with the help of good friends, that you have to put in the time to do the marketing work. For me, it's always an ongoing learning process, but I'm finally learning to treat marketing as an actual job. Whether it's a full-time or a part-time job is totally up to you. A dear buddy of mine suggested that I set aside three 2-hour work periods each week for marketing research and outreach (such as sending press packs, emails and general correspondence). It seems to be working for me ... you might give this option a try.

Distribution

In this age of digital distribution, you might be surprised that there are quite a few ways to approach the distribution of your "product". This can come in such forms as:

- Record label: yes, this is still a viable option for certain artists. In addition to the big two (Sony and Universal Music), there are a large number of independent labels out there that might be right for your music genre. Using this option, you will essentially be giving up a percentage of your gross project income and (possibly) ownership and/or distribution rights in exchange for marketing, brand notoriety and other services. As always, be careful and check things out before signing.
- Direct sales: you've all seen the street musician who's working his or her butt off on a busy street. At their feet is a CD that is for sale for $10. Or the merch table in the corner of a venue that has the band's entire CD catalog for sale, along with poster, tees and the like. These are totally viable and direct options for getting some extra income for your hard-earned work.
- Digital downloads: there are a number of digital download services that allow the fan to go on-line, listen and then buy the artist's music. This might take the form of an open system that lets you own the downloads (in MP3, FLAC or other media format), or it might be a service that still retains ownership of the actual media, but then unlocks and places it in your media library.
- Streaming: using this method, the distribution company retains all rights retain the media for distribution, allowing the user to listen to the media. In this case, there are no download options; however there are numerous options for the user to play and organize the music into personal playlists, which can then be shared.

- Music licensing for film and TV: a number of companies are available that offer on-line services for allowing motion picture, television and other media producers access to music creators on-line. This avenue can range from being modestly to quite lucrative for the artist.

The choice of any or all of these distribution services is completely up to the artist. Except for option one, the artist or producer will retain full ownership of the master and performance. With some care, this might be also available when signing with a label. My advice here would be to make your plans carefully, take your time and then make the best choice that works for you at the time.

Marketing

Once "the product" has been made, the distribution method chosen and the release date set; then comes the truly hard part ... getting it in front of the buying public and making the sale. At this point, I will not pretend to be an expert on the subject. In fact, on this subject, I'm just a beginner. However, I do know firsthand how hard this is. Even with projects that have Grammy nominations to their name, the process still takes a lot of hard work, plenty of contacts and a willingness to persevere. Here are a few questions and thoughts that you might want to ask yourself and then act upon:

- What is the initial goal that you want to set out for yourself? What is it that you want the audience to do? Check out the site? Listen to a track? Buy the album?
- Is the goal to get the product out in front of people as much as possible (free to mostly free streaming services) or to make money off of it (physical or downloadable media) ... or both?
- What is the marketing plan? Is there a website presence? Is there a social media presence? Make sure that your social media banners point to the product in some way.
- What kind of a story can be shared on social media to create a buzz? Are there videos that can be shared? Share a "making of" video or create content that can help promote the media.
- Can marketing material be gotten to a larger audience through targeted social media promotions?
- Never forget the "power of one". This refers to sending out special emails or messages to your fans, family members, past buyers, etc. You might be surprised about the impact that can be made, one contact at a time.
- Try to get your music on vlogs, blogs and sites that are related to your genre. This outreach works much like the power of one, but the "right" one just might get the word out to hundreds or thousands of new fans.
- It's easy to forget radio, but this can be a powerful medium for getting your work out to the public ... especially if they can highlight your work and the music in a more in-depth manner.

All of these things require time, research and good old-fashioned hard work. I'm in the process of learning the fine line of being persistent, without being spammy or obnoxious. Creating a schedule for getting your promotional work done is a good way to get started and to stay on track. It's not easy … but in this industry, you'll never be promised a rose garden.

Finally, there is no one right way to get your foot in whatever door you want to step through. Of course, there are a lot of books and YouTube videos on the subject of marketing. They all have their message, but in the end, you're the one who'll have to walk the long walk.

Sales

Ah yes, the big question in the arts … getting paid. Quite possibly, this is where the going gets even harder.

- For touring bands, playing live and getting direct revenue from merchandizing (merch) has become one of the best ways for artist to strive to make a living.
- Stories of problems with getting paid for artists who opt to work with labels are legendary. The contract clauses have traditionally (but not always) been written in favor of the label and distributors. Hidden fees, recouped recording costs, production costs and archaic damage and returns clauses can actually leave the artist in the hole financially. As with all things, be careful before you or your band sign your art away to another entity.
- Although downloads are becoming less popular with the masses, they can be quite straightforward with regards to getting paid. For example, my favorite way of getting my music out to my audience is through Bandcamp .com. This company allows you to sell your music in MP3 or Hi-Def FLAC (look it up on Wiki) direct to the buyer, whereby the artist or rights-holder gets 85% of the proceeds, and 90% of the proceeds from merch sales. These figures are basically unheard of in traditional music sales schemes.
- KickStarter, IndieGoGo and other crowd-funding options are another way to make money from pre-sales, even before you start work on the actual production. By getting direct support from family, friends and fans, you might be surprised how much funds you might get to support your next project.
- Streaming has definitely become the darling of the music community and for consumers for its ease-of-use and its ability to get the music out to the masses in new and innovative ways. It must be realized, however, that in this case the amount of money that even an established artist can expect to make will be modest. For example, one well-known group published an on-line story about a major release that had a total of about a million streams. From this, they received a total revenue of about $5,000, which ended up being an average per-stream payout of about $0.005. My recommendation here would be to use streaming services in ways that they work best, as a vehicle for getting the word out about your music. In short, don't buy that Tesla just yet.

STUDIO TIPS AND TRICKS

As we near the end of this book, I'd like to take some time to offer some tips that can help make a session go more smoothly, both in the project and in the professional studio environment. Almost all of these straightforward tools have to do with keeping your environment, equipment and general outlook operating at a relatively professional level. So, why would you want to have a professional approach to your production environment, even though you're working out of your bedroom? Obviously, the goal is not to make your productions sound like they were done in a dingy, old bedroom. These days, it's becoming more and more common for tracks on platinum records to be laid down in these environments. But how was the artist able to get such a professional sound? They did it by having a professional attitude towards their productions, their tools and their techniques. Armed with a positive and professional attitude, you'd be amazed at how you can step up your game. OK, let's take some time to look at some tools and techniques that can help make your projects shine.

Household Tips

Producers, musicians, audio professionals and engineers spend a great deal of time in the control room and studio. It only makes sense that this environment should be laid out in a manner that's esthetically, functionally and acoustically pleasing from a feng shui point of view. Creating a good working environment that's conducive to making good music is the goal of every professional and project studio owner. Beyond the basics of creating a well-designed facility from an acoustic and electronic standpoint, a number of basic concepts should be kept in mind when building or designing a recording facility, no matter how grand or humble. Here are a few helpful hints:

- Given the fact that an engineer spends a huge amount of time sitting on his or her bum, it's always wise to invest in both your and your clients' posture and creature comforts by having comfortable, high-quality chairs around for both the production team and the musicians (Figure 14.11).
- Velcro™ or tie-straps can be used to organize studio wiring bundles into groups that can be laid out in a way that reduces clutter, improves organization (color-coded straps can also be used) and makes the overall studio look more professional.
- Most of us are guilty of cluttering up our workspace with unused gear, papers … you know, junk! I know it's hard, but a clean, uncluttered working environment tells your clients a lot about you, your facility and your work habits.
- Keep liquids off your workspace. Just the other day, I spilled a drink on my desk and it blew out a USB-C/Thunderbolt port. Moral of the story, don't put it there in the first place … ARG!
- Unused cables, adapters and miscellaneous stuff can be sorted into plastic storage boxes and stored away to reduce clutter.
- Keep your studio in good electrical shape. A dear friend of mine (who shall totally remain nameless) had a bad habit of cobbling his cables together

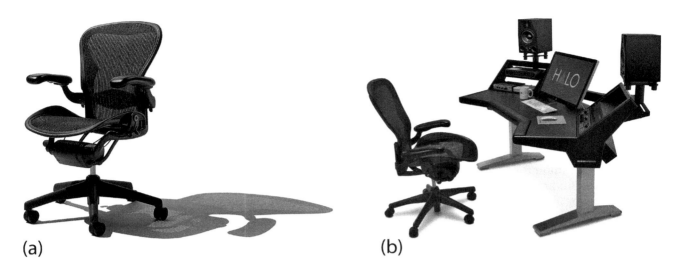

(a)

(b)

FIGURE 14.11
Functional and comfortable furniture is a must in the studio. (a) The venerable Herman Miller Aeron® chair. (Courtesy of Herman Miller, Inc.; www.hermanmiller. com.) (b) The Argosy Halo desk. (Courtesy of Argosy Console, www.argosyconsole. com.)

out of ancient connectors that had seen better days. That's all well and good, but his soldering skills weren't always the best and you never knew if the connection was going to go bad on you at any moment … and yes, good soldering skills are important to have.

- Important tools and items that are used every day (such as screwdrivers, masking tape or markers) can be stored in a rack-mounted drawer that can be easily accessed without cluttering up your space.
- Portable label printers can also be used to identify cable runs within the studio, identify patch points, I/O strip instrumentation … you name it.

Musician's Tools

By now it's probably painfully obvious to most musicians that producing the music is only the first step toward building a career in the business. It takes hard work, perseverance, blood, sweat, tears and a sense of humor (so we don't all go nuts). For every person who makes it, a large number don't. There are a lot of people waiting in line to get into what is perceived by many to be a glamorous biz. So, how do you get to the front of the line? Well, folks, here are some keys to help you on your own personal journey.

- A ton of self-motivation
- Good networking skills
- A good, positive attitude
- Realize that showing up is huge!

The business of art (the techno-arts of recording and music production being no exception) is one that's generally reserved for self-starters. Even if you get a degree from XYZ college or recording school, there's absolutely no guarantee that anyone will be knocking on the door with a job offer or contract in hand (if they do, get a lawyer, quick!). It takes a large dose of perseverance, talent and personality to make it. In fact, one of the best ways to get into the biz is to get down on your knees and knight yourself on the shoulder with a sword (figuratively or literally – it doesn't matter) and say: "I am now a _____!" Whatever title that you want to take on, just become it … Shazammm! Make up a business card, start a business, begin contacting artists to work with or make the first step toward becoming the artist you want to be. Of course, you'll need to realize that it's a step-by-step journey that can potentially last a lifetime.

There are many ways to get to the top of your own personal mountain. You could get a diploma from a school of education or the school of hard knocks (it usually ends up being from both), but the goals and the paths are up to you. Like a mentor of mine says: "Failure isn't a bad thing … but not trying is!"

Another part of the success equation lies in your ability to network with other people. Like the venerable expression says: "It's not [only] what you know … it's who you know". Maybe you have an uncle or a friend in the business or a friend who has an uncle – you just never know where help might come from next. This idea of getting to know someone who knows someone else is what makes the business world go around. Don't be afraid to put your best face forward and start meeting people. If you want to play gigs around your region (or beyond), get to know a promoter or venue manager and hang out without being too much in the way. You never know – the music maven who hangs out at your favorite café might know someone who can help get you in the proverbial door. The longer you stick with it, the more people you'll meet, thus making a bigger, stronger, better network than you ever thought would be possible.

Like my own music maven always says, "Showing up is huge!" It's the wise person who realizes that being in the right place at the right time means being at the wrong place hundreds, if not thousands of times. You just never know when lightning is going to strike – just try to be standing in the right field when it does.

Here are some more practical and immediate tips for musicians:

- Build a personal and/or band website: making your own personal site will help to keep the world informed of your gigs, projects, bio and general goings-on.
- Of course, you're aware that getting your music distributed is super-easy and very affordable or free. Services like Distrokid can get your music out to the various on-line download distributors.
- Build a relationship with a music lawyer: many music lawyers are open to building relations that can be kicked into gear at a future time. Take the time to find a solicitor who's right for you. Does he or she understand your personal music style? If you don't have the bucks, is this person willing to work with you and your budget, as your career grows?
- The same questions might be asked of a potential manager, although this symbiotic relationship should be built with care, honesty and safeguards (just one of the many reasons you might want to know a music lawyer).

- Copyright your music: as was said earlier in this chapter, it's often a good idea to protect your music by registering it with the Library of Congress or more simply through the simple use of "Creative Commons". It's easy and inexpensive and can give you peace of mind about knowing that the artistic property that you're sending out into the world is protected.

A WORD ON PROFESSIONALISM

Another subject that should be touched upon is the importance of a healthy professional demeanor, even in your own project studio. Without a doubt, the life and job of a producer or musician aren't always easy. The work often involves long hours and extended concentration with people who, more often than not, are new acquaintances. In short, it can be a difficult to remain focused on both your music and your business. On the flip side, it's one that's often full of new experiences and it lets you be involved with people who feel passionately about their art and chosen profession.

It's been my observation (and that of many I've known) that the best qualities that can be exhibited by anyone in the biz are:

- An open heart
- An innate willingness to experiment
- Openness to new ideas (flexibility)
- An openness to admitting when you're wrong or have made a mistake
- A sense of humor
- Good personal grooming habits
- An even temperament (this often translates as patience)
- A willingness to communicate openly and clearly with others
- An ability to convey and understand the basic nuances of people from all walks of life

The best advice I can possibly give is to be open, be patient and, above all … "be yourself". In addition, be extra patient with yourself, if you don't know something … ask. If you make a mistake (and trust me, you will – we all do), admit it and don't be hard on yourself. It's all part of the process of learning and gaining experience.

Choosing Your Gear

Choose the best gear that you can, given your budget. There are so many types, manufacturers and quality levels for gear out there that it's difficult to make the best choice. What I can say is take your time, don't listen to marketing hype and find the best gear that will work for you … at a price that you can afford. The one thing that I have learned over the years is that everything has its own sound. This also goes for digital gear. For example, different audio interface models will always sound different. They all have their own character. Obviously, the same goes for speakers … and these two are probably your most important hardware choices. My advice here is to take your time in buying, do as much research as possible and buy from a store that will offer returns or exchanges (if possible). Good luck!

DIY

Here, I'd like to take a moment out to put in a word for self-sufficiency and saving your hard-earned cash. If you have an absorber that needs to be built, you might try building it yourself. If you need a set of cables that need to be assembled, instead of rushing out and paying a small fortune for them, you might consider getting what you need (if you don't have something close already) and putting them together yourself. In fact, I'm going to challenge each of you to learn how to solder. Learning this craft will come in handy throughout your career. If a cable breaks, instead of throwing it out, you could fix it yourself. In addition to soldering, a good understanding of safe and basic construction and electrical contracting skills will also go a long way towards your being able to get the job done, at a fraction of the cost and with twice the amount of personal pride. Go ahead … get a decent soldering iron and start making or fixing some cables for practice. I personally think that basic DIY skills are necessary for any self-respecting project studio owner. Have fun, visit YouTube, be safe and keep your quality control high.

Protect Your Investment

When you've spent several years amassing your studio through hard-earned sweat-equity and bucks, it's only natural that you'll want to take the necessary precautions to protect your investment. Obviously, the first line of defense is to protect your data. This is done through a rigorous and straightforward backup scheme (again, the general rule is that something isn't backed up unless it's saved in three places – preferably with one of the backups being stored off-site).

The next step in taking care of your studio is to protect your hardware and software investments as well, by making sure that they're properly insured. The best way to start this process is by contacting your trusted insurance agent or broker. If you don't have one, now's the time to get to know one. You might get some referrals from friends or people in your area and give them a call, set up some appointments and get several quotes.

If you haven't already done so, sit down and begin listing your equipment, their serial numbers and replacement values. Next you might consider taking pictures or a home movie of your listed studio assets. These steps will help your agent come up with an adequate replacement plan and will come in handy when filing a claim, should an unfortunate event occur. Being prepared isn't just for the Boy or Girl Scouts.

Taking Care of Your Most Valuable Investment ... You!

When buying equipment, make sure to take a breath and assess whether or not your pocketbook can handle the expense of that latest and greatest new toy. Will it fit into your budget? Can it be written off as a tax expense (it's always a good idea to make your business work for you, come tax time)? How can I recoup the cost within my current business plan?

In addition to all of the business and household expenses, I'd urge you (whenever possible) to open a low-to-no-fee savings account with your private banker (if you don't have one, visit your bank and get to know him/her ... this always comes in handy in a pinch). The fact is, as an artist, you almost certainly won't be assured a steady income. For this reason alone, it's critical that you deal with your business and personal finances responsibly. If you don't know how to start, sit down and develop a plan with your newly found personal banker.

Update Your Software

Periodic software updates might help to solve some of those annoying problems that you've been dealing with in the studio. Many times the software that's been pressed onto a CD or DVD that came in the box will be out of date by the time it reaches you. For this reason, it's a good idea to check the Web regularly to see if there's a newer update version that can be loaded.

A word of caution, not all updates will be your friend. As with all things, Murphy's Law can creep into an update that can cause problems. Before doing an important update (although they all potentially are), you might consult the Web to see if others have had issues before you. Additionally, never do an update just before an important session or just before hopping on a plane; it's no fun being in a critical situation only to find that your update just crashed the system.

Read, Read, Read...

As electronic musicians, it's important that we stay on top of the new gear, the latest technological advances and new/old production techniques. How can we best do that? For starters, you can get the best results from your gear by reading your manuals. For those of us who hate reading manuals (and I head that list), get on the Web and ask your fave browser a question: Google? How long can a Thunderbolt cable be to run at 40 Gb/s? Or how do I connect my wireless controller app to my DAW?

In addition to the Web, various recording and electronic music mags (available both for free and through paid subscription) can be placed in the bathroom as reading material (something that's always been a personal joy for me). Books can come in really useful. Last, but definitely not least: go to industry-related conferences. It's one of the smartest things you can do to broaden your network beyond your back yard! Keeping on top of the tools, toys and techniques can definitely be fun! Get out there and grab yourself some knowledge that you can put into practice!

IN CONCLUSION

Obviously, the above lists and thoughts are just the beginning of an ever-changing set of tips that can help. The process of producing, recording and mixing in any type of studio environment is an ongoing, lifelong pursuit. Just when you think you've gotten it down, the technology or the nature of the project changes right under your nose and hopefully, you'll learn from this never-ending educational journey. Far more than just the technology, the process of coming up with your own production style and your own way of applying the tools, toys and techniques to a production is what makes us artists, whether you're in front of the proverbial glass and/or behind it. Over time, your own list of studio tips and tricks will grow. Take the time to write them down and pass them on to others on your Web blog and be open to taking the advice of your friends and colleagues. Use the trade mags, conventions and the Web to open yourself up to new insights, to better use the tools of your profession and to find new ways of doing "stuff". Learning is an ongoing, ever-changing process and most importantly: have fun along the way!

Hugs,
DMH

Index

Page numbers in *italics* indicate figures. Page numbers in **bold** indicate tables.